Fluxgate Magnetometers for Space Research

editor

Günter Musmann

January 26, 2010

Bibliografische Information der Deutschen Nationalbibliothek
Die Deutsche Nationalbibliothek verzeichnet diese Publikation in der Deutschen Nationalbibliografie;detaillierte bibliografische Daten sind im Internet über dnb.d-nb.de abrufbar
Fluxgate Magnetometers for Space Research / Günter Musmann (Editor). – Salzgitter,

Herstellung und Verlag:
© Books on Demand GmbH, Norderstedt
ISBN 978-3-8391-3702-4

All rights reserved (including those of translation into other languages). No part of this book may be reproduced in any form – by photoprint, microfilm, or any other means – nor transmitted or translated into a machine language without the written permission from the editor.
Registered names, trademarks, etc. used in this book, even when not specifically marked as such, are not to be considered unprotected by law.

Composition: L. Kastrup, G. Musmann,

Fluxgate Magnetometers
for Space Research

Preface 1

Part I of this book is devoted to the theory and practice of fluxgate magnetometers.
Yuri Afanassiev, the author of Part I wishes to thank Günter Musmann, for his initiative, friendly approach and support for the translation first into German and later also into English. As a result of Musmann's initiative, this book forms a two parts comprehensive work. The second part involves worldwide team work, and is based on achievements in fluxgate magnetometer design of the past decades.
The authors are grateful to G. Musmann's colleagues M. Rahm and I. Richter for the constructive criticism of some parts of the book.
The authors also thank Elena N. Tschishowa, who has undertaken the translation of the book and to T. Tittmann who helped with physical wording and translation and to E. Serpell (formerly Imperial College, London) who corrected and reworded the first draft version.
The authors hope that this book will be of use as well to experts as to students working in the field of designing and applying fluxgate magnetometers.

In Memoriam:
Mario H. Acuña and Robert A. Langel
All of us had the good fortune to have worked with these two charming, powerful and thoughtful men who were full of energy, clear thinking, genuine interest and curiosity and having an unfailing willingness to achieve the best possible results. We had the pleasure of working with them on space magnetic field experiment projects for years , they will always be remembered.

Günter Musmann, Yuri V. Afanassiev
Salzgitter, Germany
2009

Preface2

Scientific and technological progress is closely linked to the improvement of existing apparatus and development of new, more sensitive and accurate instruments.

Among these instruments, an important place is occupied by instruments designed for measuring parameters of magnetic fields. These are used in prospecting mineral resources (oil, gas and condensate), studying the space and time structure of the magnetic fields of the Earth and planets, orienting and determining a locus of different objects moving in the Earth's field, controlling quality of produced electrotechnical steels, different materials and products, and for some other objectives. Most generally used are the fluxgate instruments, the characteristics of which include not only high sensitivity and accuracy, but also the ability to directly measure the components of a magnetic field (thereby providing complete information on the field structure and field sources), and suitability for measuring very weak magnetic fields, in a wide temperature range and in the presence of electromagnetic noise. Other features of these instruments are high safety, longevity and low cost.

The achievements in fluxgate instrument production over the last about 40 years are quite considerable. These advances are due to development of automated experimental measurement techniques and radio electronics, as well as to advances in physics, metallurgy and fabrication of materials.

Intrinsic noise in fluxgate instruments are now of the order of 10^{-12} THz$^{-1/2}$, 10 – 100 times lower than their level achieved earlier. This allows fluxgate instruments to be employed not only in the traditional way but also in new directions. The latter involve studying the remanent magnetisation of slightly magnetic rocks (paleomagnetic researches), recording magnetic disturbances caused by physico-chemical processes in biological systems (including the human cardio-vascular system), uncovering micropulsations of the electromagnetic field in the zone of a so-called planetary shock front and in interplanetary space etc. Then accuracy of fluxgate instruments has been increased too. Vector field compensation inside a fluxgate volume and transformation of compensation currents into digital codes have given rise to the development of high-speed wide-range instruments of a new generation, the relative accuracy of which amounts 0.01% or smaller.

This book is not intended as a reference guide which describes the circuits, constructions and characteristics of all fluxgate instruments both in production or existing as unique prototypes. Examples of individual instruments are presented in the book, this is made solely to illustrate trends and ideas at the present stage of development of fluxgate techniques.

The contents of the book are dictated by its objectives which are: the theoretical basis and presentation of the trends and ideas aimed at further improvement of fluxgate instruments.

The book is comprised of two parts. The first is devoted to the basics of the design theory of fluxgate units, the properties and parameters of which determine, as a rule, the instrument characteristics. Considerable attention is given to calculating fluxgate elements, comparing their parameters, revealing sources of zero displacements and noise, and choosing ferromagnetic and other materials needed for constructing the instruments. The advantages of fluxgate construction with closed cores (over those with open cores used earlier) by using amorphous materials are emphasized. The first part of this book can be considered as an addition to the monograph "Fluxgate magnetometers" issued by Y.Afanassiev , "Energiya" in 1969.[1]

The second part of the book is devoted to the particular features in designing modern fluxgate instruments for space research. Attention is given to functional design and sources of error for the instruments. Detailed descriptions are provided for instruments with high-precision wide-range instruments (including digital ones). Means of certification ,calibration on ground and inflight and test for high and low temperature of these instruments are briefly described. Because high resolution magnetic field measurements in space require a magnetic cleanliness program for the spacecraft a separate chapter (Appendix) is added on - Guidelines for Magnetic Cleanliness-.

---

[1] Yuri Afanassiev,the author of part I, expresses his gratitude to the Scientific Councils of the Institute of Cosmic Investigations,the D.I.Mendeleev Institute for Metrology, AN SSSR, and to the Research-and-Production Association "Rudgeofizika", which realized the necessity to issue this book and corrected its contents. The author is also thankful to his colleagues V. I. Sheremet, V. Ya. Shifrin, V. A. Prishchepo, V. P. Porfirov, Yu. N. Bobkov, and others for the useful discussions and help in writing part I of the book.

Yuri V. Afanassjev

Fluxgate Magnetometers for Space Research

Part I – Basic Theory

by Yuri V. Afanassiev

Translation from Russian by Elena Tschishowa
Reviewed by Ed Serpell

Part II – Applications

Contributions from

Mario Acuña, Yuri V. Afanassiev, André Balogh,
Chris Carr, Karl H. Fornaçon, Robert A. Langel,
Werner Magnes, Karl Mocnik, Günter Musmann,
Holger Kügler, Falko Kuhnke, Hermann Lühr,
Otto V. Nielsen, Fritz Primdahl, Terence J. Sabaka
and Robert C. Snare,

Contents

1 **Design and Operation Principles**   3
   1.1 Historical Overview 3
     1.1.1 Basic Definitions 3
     1.1.2 Invention of Fluxgate Magnetometers 4
     1.1.3 Aeromagnetics Forcing the Development 7
     1.1.4 Space Applications 8
     1.1.5 History of Theory Development 9
   1.2 The Fluxgate as a Magnetic Modulator 10
     1.2.1 The Fluxgate with Parallel Fields 11
     1.2.2 The Fluxgate with Perpendicular Fields 12
     1.2.3 Magnetic Modulation 13
     1.2.4 Two Basic Operation Modes 14
   1.3 Ferromagnetic Materials Characteristics 16
     1.3.1 From Hysteresis Loops to Mean Magnetization Curves 18
     1.3.2 Normal and Differential Permeabilities of Matter . . . 19
   1.4 Demagnetization Factor. Permeability of Cores. 21
     1.4.1 Tensor of a Body's Permeability 21
     1.4.2 Normal and Differential Permeabilities of a Body . . . 22
     1.4.3 Autostabilization of Body Permeability 25
     1.4.4 Calculation of the Permeability of Ellipsoidal Cores . . 27
     1.4.5 Calculation of the Permeability of Non-ellipsoidal Cores 28

2 **Fundamentals of Parametric Theory**   32
   2.1 Magnetic Quantities and Parameters 32
   2.2 The Influence upon Permeability 36
     2.2.1 Transversal Excitation 36
     2.2.2 Longitudinal Excitation 37
     2.2.3 Auto-Parametric Approach 39
   2.3 Derivation of the Most General Expression for the Output
     EMF 40
     2.3.1 Tensor of the Dynamic Permeability of a Body 42

		2.3.2	Special Cases of Excitation	43
		2.3.3	Calculation of the Transformation Ratio	45
		2.3.4	Calculation of the Misbalance EMF	47
		2.3.5	A General Expression for the Output EMF	48
	2.4	Open Core Fluxgates		50
		2.4.1	Voltage Mode Excitation	53
		2.4.2	Additional Data for Transformation Ratio Calculation	56
	2.5	Closed Core Fluxgates		57
		2.5.1	Types of Closed Core Fluxgates	57
		2.5.2	Notes for Deriving Formulae	59
		2.5.3	Recommended Method of Estimation	60
		2.5.4	Extension of the Method to Transversal Excitation . .	63
	2.6	Modes of Inner Parametric Amplification		64
		2.6.1	Conditions for Measuring Circuit Stabilization	65
		2.6.2	Conditions for Excitation Circuit Stabilization	68
	2.7	Magnetic Axis of Fluxgates		70
		2.7.1	The Concept of the "Magnetic Axis"	70
		2.7.2	Causes of Destabilization of the Magnetic Axis	71

3 Magnetic Offsets and Noise   73

3.1	General Statements			73
3.2	Mechanisms of Magnetic Shifts			75
	3.2.1	Mismatch of the Fluxgate Half Cells		76
	3.2.2	Remanent Magnetization		77
	3.2.3	Magnetostriction Effect		78
3.3	Magnetic Noise			82
	3.3.1	Modelling Magnetic Noise		84
3.4	Effects of the Drive Field			90
3.5	Noise, Magnetostriction, Magnetic Anisotropy			97
3.6	Other Relations and Dependencies			104
	3.6.1	The Dependence on the Drive Field Frequency		104
	3.6.2	Operational Temperature Dependence		105
	3.6.3	Spatial and Temporal Temperature Variations		106
	3.6.4	Volume, Shape and Processing Effects		106
3.7	Prospects for Suppressing Noise			108
	3.7.1	Magnetic Material Parameter Selection		109
	3.7.2	Curie Point Dependence		109
	3.7.3	Amorphous Materials		111

4 Functional Design and Error Calculation — 113
- 4.1 Functional Design 113
  - 4.1.1 Magnetometers for Constant Fields 113
  - 4.1.2 Magnetometers for Alternating Fields 114
  - 4.1.3 Magnetic Gradiometers 116
- 4.2 Total Error 116
- 4.3 Methods for the Minimization of Multiplicative Errors 120
  - 4.3.1 Errors of Transformation Ratio 120
  - 4.3.2 Angular Errors 122
  - 4.3.3 Errors Caused by Feedback Coils 125
- 4.4 Means for Additive Error Minimization 127
- 4.5 References chapter 1-4 130

5 General Fluxgate Magnetometer Design — 137
- 5.1 Instrument Design 137
  - 5.1.1 General 137
  - 5.1.2 The Ring Core Fluxgate 138
  - 5.1.3 The Drive Generator 143
  - 5.1.4 The Short-Circuited Fluxgate Sensor 146

6 Magnetic Materials for Sensors — 152
- 6.1 Common Requirements for the Magnetic Materials 152
- 6.2 Crystalline Materials 154
  - 6.2.1 Amorphous Materials (Metallic Glasses) 158

7 Magnetic Field Vector Feedback — 169
- 7.1 Introduction 169
  - 7.1.1 Triaxial Feedback Configuration 171
  - 7.1.2 Vector Feedback Systems 173
  - 7.1.3 Stability of the Vector Feedback System 174

8 Digital Magnetometer — 177
- 8.1 Digital Detection and Feedback Fluxgate Magnetometry ... 177
  - 8.1.1 Introduction 177
  - 8.1.2 The B-Field Related Sensor Output Signal 178
  - 8.1.3 The Impulse Integration Technique 179
  - 8.1.4 Real-Time Second Harmonic Digital Detection and Feedback Magnetometer 181
  - 8.1.5 Digital Cross-Correlation Detection and Feedback Fluxgate Magnetometer 183
  - 8.1.6 The Detection and Control Algorithms 185

	8.1.7	The Astrid-2 Satellite Magnetometer	187
	8.1.8	Discussion and Conclusions	188

9 Magnetometer Ground Calibration   191
9.1 Calibration Principle 191
9.2 Co-Ordinate Systems 192
9.3 Simple Example 193
9.4 Magnetic Environment Requirements During Calibration . . . 197
9.5 Parameters of a Linear Instrument 198
9.6 Modelling Non Linear Effects 205
    9.6.1 Assumption of a Non Linear Sensitivity 205
    9.6.2 Assumption of a Non Linear Dependency of Orientation 206
    9.6.3 Model without any Pre-Knowledge of the Transfer Function 207
9.7 Determination of Linear Model Coefficients 207
9.8 Estimation of the Quality of Fit 208
9.9 Offset of Magnetometer and Residual Field of Facility 209
    9.9.1 Determination of Facility Offset and Residual Field . 210
    9.9.2 Influence Elimination of Offset and Residual Field . . 210
9.10 Transfer Function, Influence of Systematic Errors 211
9.11 Selection of Suitable Test Vectors 211
    9.11.1 On-Axis Measurements 212
    9.11.2 Spike–Sphere–Measurement 213

10 Magnetometer Inflight Calibration   216
10.1 Non-Absolute Satellite Vector Magnetic Field Data 217
    10.1.1 Introduction 217
    10.1.2 Calibration of a Vector Magnetometer with Absolute Scalar Data 220
    10.1.3 Calibration of Magnetometer Attitude 227
    10.1.4 Discussion 233

11 Temperature Test Facility for Magnetometers   236
11.1 Introduction 236
    11.1.1 Magnetic Shielding Set 237
    11.1.2 Low Temperature Equipment 237
    11.1.3 High Temperature Equipment 237
    11.1.4 Calibration Coil 239
    11.1.5 Test Result 239

12 General Fluxgate Magnetometer Bibliography   242

13 APPENDIX
Guidelines for Magnetic Cleanliness   254

- 13.1 Introduction 256
- 13.2 Design Approach 258
  - 13.2.1 General 258
  - 13.2.2 Approach to Minimizing Fields from Magnetic Materials 258
  - 13.2.3 Approach to Minimizing Stray Fields from Electrical Currents 259
  - 13.2.4 Approach to Minimizing Fields from Eddy Currents . . 259
  - 13.2.5 Magnetic Moment Management , Approach to Hardware Analysis 259
  - 13.2.6 Shielding Problems 260
  - 13.2.7 The Approval Process 261
- 13.3 Achievement of Magnetic Cleanliness 261
  - 13.3.1 Permanent Fields and Fields from Soft Magnetic Materials 262
  - 13.3.2 Stray Fields Produced by Electric Currents 266
- 13.4 Magnetic testing of Spacecraft: Flight Hardware 271
  - 13.4.1 Testing at Assembly and Unit Level 271
- 13.5 Appendix A 275
  - 13.5.1 Relations to other units : 275
- 13.6 Appendix B 276
  - 13.6.1 GENERAL GUIDELINES FOR THE DESIGN OF MAGNETIC CIRCUIT FOR MAGNETICALLY CLEAN SPACECRAFT 276

Part I - Basic Theory

Chapter 1

Design and Operation Principles

Yuri Afanassiev

Mag-Sensors, St.Petersburg

1.1 Historical Overview

1.1.1 Basic Definitions

Fluxgates are devices, sensitive to the influence of external magnetic fields, which consist of ferromagnetic cores wound with two coils, to one of which alternating current is fed, and on the other one an electromotive force EMF, proportional to the external field, is generated. Fluxgates are related to induction transformers.

The difference between passive induction transformers, including ferroantennas, and fluxgates is that fluxgates are induction transformers of the active type. Transformation processes, taking place within them, are always connected with the existence of at least two magnetic fields: the external (as a rule, constant or slowly changing) field and the auxiliary or driving variable field, created by alternating current. The counteraction of these fields in the volume of ferromagnetic cores results in an EMF across the sense coil, which contains information about the external field.

The operation of a fluxgate is closely related to that of magnetic amplifiers, which was known long before fluxgates were invented [36, 56]. Actually, the fluxgate is a kind of magnetic amplifier. In Fig. 1.1 you can see the arrangement of a magnetic amplifier and a fluxgate. In a magnetic amplifier (Fig. 1.1a) the current i_0, feeding the incoming winding, first transforms in the strength of the circular

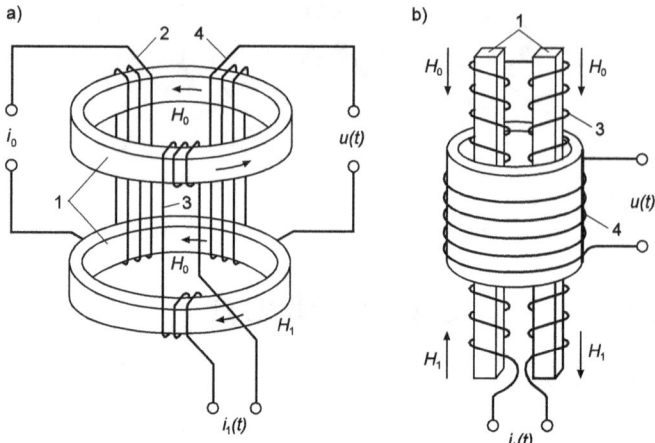

Figure 1.1: The arrangement of a magnetic amplifier (a) and a fluxgate (b): 1 – ferromagnetic cores; 2 – the input winding; 3 – the alternating current winding; 4 – the output measuring winding.

magnetic field $H_0(i_0)$ and then, together with the strength of the alternating field $H_1(t)$, created by the current $i_1(t)$ transforms in the ferromagnetic cores into variable induction $B(t) = B[H_1(t), H_0]$ and the EMF $u(t) = u[B(t)]$ in the outcoming winding of the amplifier.

In a fluxgate (Fig. 1.1b) the external field with the strength H_0 directly influences the ferromagnetic cores. The alternating field $H_1(t)$, created by the current $i_1(t)$, also influences the cores. The transformation of the fields H_0 and $H_1(t)$ into alternating magnetic induction $B(t) = B[H_1(t), H_0]$ and afterwards, into the EMF $u(t) = u[B(t)]$, appearing at the sense coil, is performed in a similar way to that of magnetic amplifier.

Fluxgates differ from magnetic amplifiers because they lack the incoming winding. Instead of it, fluxgates contain ferromagnetic cores, which are magnetized by the applied field. The sense coil is wound on these cores such that the EMF $u(t)$ induced is a function of the strength H_0 of the external field to be measured.

1.1.2 Invention of Fluxgate Magnetometers

In Russian and foreign works the invention of fluxgates is associated with the German scientists H. Aschenbrenner and G. Goubau [51].

The fluxgate of Aschenbrenner and Goubau is shown in Fig. 1.2. The ring core is made of iron wire with the diameter of 0.2 mm, isolated with shellac. The drive coil is wound around the core. The sense coil is wound on a framework, enveloping the ring core. A 500 Hz current $i_1(t)$ feeds the drive coil. The output

1.1. HISTORICAL OVERVIEW

EMF of 1000 Hz, which carries information on the strength of the field $H_{0,i}$ with $i = x, y, z$, lying in the plane of the ring core and aligned with the normal to the plane of the sense coil, appears across the sense coil. To increase sensitivity near the ring core ferromagnetic plates, acting as concentrators of the external field (not shown in Fig. 1.2), are used. The device consists of signal generators, an amplifier, an asynchronous detector, a direct current amplifier and a display, besides the fluxgate and field concentrators. Using this device the inventors discovered a zero offset. The reasons for the zero offset were not found; the device was recommended mostly for measuring magnetic fields of low frequency, and for the recording of short period variations of the magnetic field of the Earth (the geomagnetic field) in particular.

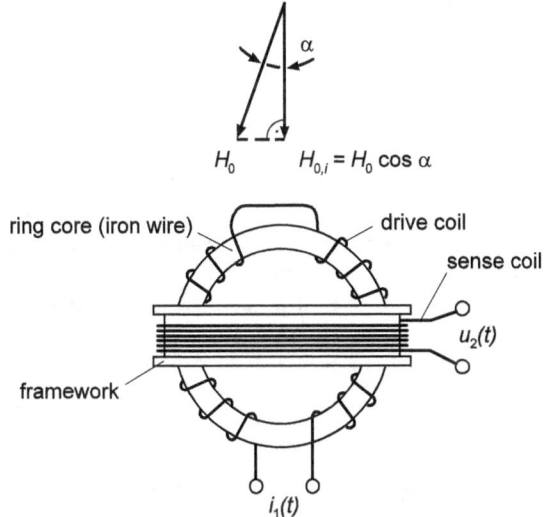

Figure 1.2: The Aschenbrenner and Goubau fluxgate.

Later F. Förster suggested using rod cores instead of ring cores in fluxgates. Förster used two identical parallel rod cores, placing not only the drive coil, but also the sense coil on each of them (see Fig. 1.3). This made possible to make use of the fluxgate as both a transformer of the full field strength (Fig. 1.3a), and a transformer of the field strength difference. The switch from one variant to the other was carried out only by turning one of the semi-elements of the fluxgate by 180°. Using double rod fluxgates Förster designed a number of devices including those for the US Mariner Aircrafts, which have been produced by the company "Institut Dr. Förster" (Reutlingen, Germany). Independently of Förster's works, approximately at the same time, rod fluxgates were developed in Russia. For example, the Halileyev fluxgate [3] contains one core and one winding, performing

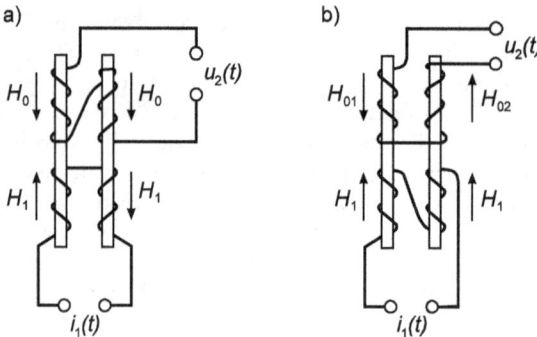

Figure 1.3: The Förster fluxgate as a field total sensor (a) and a field difference sensor (b).

the role of both the sense and the drive windings. At the beginning of the forties the Halileyev fluxgate was used for prospecting of iron ores.

Other important research was done by the famous Soviet physicist G. S. Gorelick [15] and reference should also be made to the works of E. Harrison. Harrison considered the dependence of the resistance of ferromagnetic wire, fed with weak alternating current, on the external field, influencing it in the longitudinal direction. He found that the effect is a great deal more significant than the magnetic-resistance (changing of the resistance under the influence of the perpendicular magnetic field). In fact, Harrison studied the effect inverse to that discovered by Soviet scientists.

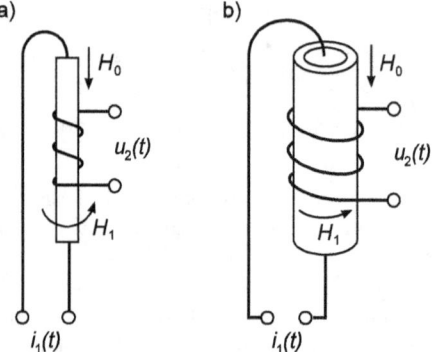

Figure 1.4: Fluxgates with perpendicular fields: a) with a wire core; b) with a tube core.

Soviet scientists invented a fluxgate with perpendicular fields at the Gorky

1.1. HISTORICAL OVERVIEW

Physico-Technical Research Institute [7, 15, 60] in 1944. The fluxgate contained a rod (wire) core, fed with a relatively strong drive current. The sense coil was wound around the core (Fig. 1.4a). In case of a constant external field H_0, a second harmonic EMF in the sense coil was created. The mechanism of the EMF creation is not trivial. They found the dependence of the longitudinal component of induction, caused by the external field, on the intensity of the orthogonal field, caused by the drive current. This is an effect similar to that studied by Harrison. The authors not only demonstrated these effects, but explained both effects and helped to develop fluxgate theory. The best parameters of fluxgates with perpendicular fields were discovered after replacing wire cores with tubular ones [3, 63, 64]. The tubular fluxgate scheme is shown in Fig. 1.4b.

1.1.3 Aeromagnetics Forcing the Development

Intensive study of fluxgates and fluxgate devices started with the exploration and application of aeromagnetic surveying.

This method was started in Russia by A. A. Logachov using an induction magnetometer with a rotating multi-curve coil [24]. The magnetometer was fixed on an aeroplane, the first experimental flight taking place in 1936. Due to its efficiency this method soon became one of the main ones in carrying out magnetic prospecting work and geological charting.

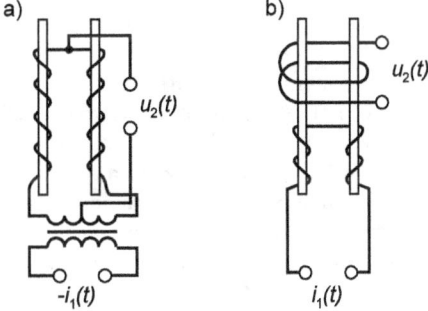

Figure 1.5: Fluxgates, used in aeromagnetometers: a) with matching windings; b) with separate windings.

Not long before the Second World War, V. Vaquet drew the attention of American scientists to German works on fluxgates. Americans started carrying out research on a grand scale, taking into consideration different variants of fluxgates, until they focussed their attention on the design shown in Fig. 1.5. They designed improved electronic contacts, used synchronous detecting and field compensation for improved accuracy. Americans invented a method of automatically orientating the fluxgate in the direction of the magnetic field vector by using two other fluxgates included into the system of the driving gear. This invention proved to be important, making it possible to measure the scalar value of the field, the vector modulus of the geomagnetic field, and to achieve (for those times) an extremely high sensitivity while carrying out measurements in motion. By the beginning of the forties Americans had already fluxgate aeromagnetometers designed to search for submarines. Later, devices suitable for geological research and for oil and gas prospecting, were developed. Publications on these devices appeared only at the end of the 1940s [47, 48][1].

In Russia the first models of fluxgate aeromagnetometers were designed under the supervision of G. S. Smirnov [3]. At the end of the fifties more accurate magnetometers were designed, which were later produced in series [3]. These made it possible to carry out the magnetic survey of vast regions and defined water areas of the country on a grand scale, resulting in discoveries of new areas of oil, gas, rare metals and other minerals. The survey results are still used for carrying out tectonic research, studying the geological structure of the ocean bottom and making oil and gas area discoveries.

It is important to mention devices constructed with pendular or gyroscopic orientation of fluxgates. In the United Kingdom an aeromagnetometer for measuring the geomagnetic field scalar and magnetic declination and inclination was made. In Canada under the supervision of P. Serson [70] a "flying laboratory", containing a three-component magnetometer with gyroscopic fluxgate orientation, data acqusition and data processing was made. In the Soviet Union Sh. Sh. Dolginov created a three-component magnetometer, later used on the "Zary" schooner. Brief information on these devices can be found in [3].

Three-component magnetometer devices with fluxgate orientation have not been widely used because they are too complicated and because of the considerable errors of such orientation systems due to small changes of the field being measured. As it has already been mentioned, the best measurements in motion have been made using scalar devices. Scalar fluxgate aeromagnetometers were widely used for mineral searches and geographical survey until they were replaced by, also scalar but more accurate, quantum (nuclear- and atomic-processing) magnetometers.

1.1.4 Space Applications

A new stimulus for the development of fluxgate devices came with the beginning of space exploration. It coincided with a revolution in electronics, the transition from vacuum tubes to semi-conductors, thus decreasing overall dimensions, weight and power consumptions of devices and increasing their reliability.

On the third Soviet satellite, "Sputnik 3" (1958), a fluxgate magnetometer, made under the supervision of Sh. Sh. Dolginov [18], was used and functioning

---
[1] See also lists of literature given in [3, 35, 36, 62, 63, 64]

correctly. In functional and kinematic design the device resembled an aeromagnetometer, but besides the field magnitude it measured two other angles, defining the field vector in the coordinate system of the satellite. Variable resistors, kinematically connected with the elements of the driving gear of the magnetometer, were used as the angle sensing elements. It was possible to measure the position of the satellite in reference to the geomagnetic field vector and its spin. The electronics of the device were designed with transistors.

The experiment which was carried out [18], gave birth to magnetic measurement in space.

In the sixties in the USSR, the USA and Germany three-component (unoriented) fluxgate magnetometers, characterized by high sensitivity and resolution [2, 3, 62, 78], were invented. These devices were used in missions which discovered and studied the magnetic fields of the Moon, Venus, Mars, and other planets. Devices of this kind are widely used in space exploration to the present days.

Satellite magnetometers required a considerable improvement of metrological characteristics. First of all, it was necessary to decrease the zero offset and the noise level of fluxgates themselves.

Analyzing different constructions of fluxgates, R. Ya. Berkman in the USSR [5][2] and V. Yager in the USA [56] independently arrived at the conclusion that fluxgates with ring cores would possess the best characteristics.

D. Gordon (USA) [57, 59] has persuasively proved this. Having actually returned to the Aschenbrenner and Goubau design (see Fig. 1.2), Gordon improved upon its implementation. He used a new ferromagnetic material, which, shaped in a narrow thin tape, was wound on a non-magnetic metal bobbin, the coefficient of thermal expansion of the tape and the bobbin being approximately the same. The core, made in this way, was subjected to heat treatment, then with toroidal drive coil on, it was put inside the sense coil cavity, where it was free to rotate. Due to this rotation fluxgate balancing could be fulfilled (the misbalance EMF, created by the mismatch of the core halves, as well as the zero offsets, were reduced). In comparison to rod fluxgates, used before, the Gordon ring fluxgate had a considerably lower noise level, which recommended its use in space magnetometers.

Nowadays, the advantages of fluxgates with ring and other closed cores are widely recognized. In the last 20 years fluxgate devices, characterized not only by a low noise level (a low sensitivity threshold), but also greater accuracy have been developed.

1.1.5 History of Theory Development

Though ferroprobes were invented relatively long ago, a general and consistent theory of fluxgates, which might serve the basis of analyzes, calculation and design of fluxgate devices, appeared considerably later.

Soviet scientists played an important role in the fluxgate theory development.

---

[2]The ring fluxgate with low noise level was suggested by R. Ya. Berkman in 1959.

First of all, it is necessary to mention the theory of solid bodies of finite sizes magnetized in an external field, which had been developed by famous physicist V. K. Arkadyev [1] long before fluxgates were invented. Laying a definite borderline between permeability of the substance, the shape and the solid body, V. K. Arkadyev wrote the expressions, which became basics of fluxgate theory.

On the basis of this theory M. A. Rosenblat suggested, and proved by experiment, the universal semi-empirical expression for calculations of the shape permeability of rod cores demagnetizing coefficient [34]. Rosenblat also gave formulas for engineering design of rod fluxgates [35, 36].

Still earlier G. S. Gorelick developed the theory of fluxgates with perpendicular fields [15]. The parametrical essence of the transformation processes has been pointed out in this work rather clearly.

A general theory of fluxgates with parallel fields was developed by R. I. Yanus, L. H. Freedman and V. I. Drozhzhina. Not aiming at any particular approximation of the rod cores magnetizing contour, using the expansion of the nonlinear dependence in the Taylor series, the authors got the expression for the peak, average and amplitude values of the output EMF of the fluxgate. This work served as a correct introduction to fluxgate parametric theory (see Part II). Before it, parametrical views were either ignored [35, 47, 51], or simplified [71][3].

Further development of the fluxgate parametrical theory was stimulated by works of Yu. F. Ponomaryov, L. Ya. Mizyuck [28], R. Ya. Berkman [5, 6].

In 1969 the "Ferrosondes" monograph was published. It was the first general work, in which fluxgate theory, design and application questions were discussed. Special emphasis was placed on the correct nature of the parametrical theory and its usefulness for the purpose of describing transformation processes in ferroprobes of different types, and also for studying similar phenomena and effects in them.

The modern stage of the fluxgate parametric theory development is characterized by containing not only the processes of transformation of the field into the EMF, but also processes relating to fake signal and noise generation.

1.2 The Fluxgate as a Magnetic Modulator

As mentioned in Sect. 1.1.1 the action of the fluxgate is connected with the presence of at least two magnetic fields – the external field to be measured and the auxiliary field, created by the drive current.

In general, the superposition of these fields is given in the expression

$$\mathbf{H}_\Sigma(t) = \mathbf{H}_0 + \mathbf{H}_1(t) \tag{1.1}$$

where $\mathbf{H}_0$ is the strength of the external field, $\mathbf{H}_1(t)$ is the strength of the drive field.

---

[3]It was also Yanus, Freedman and Drozhzhina, who suggested the term "Ferrosonde", which is most widely used in Russian literature.

1.2. THE FLUXGATE AS A MAGNETIC MODULATOR

The magnetic induction vector **B**, inside homogeneously magnetized cores, can be found on the basis of the dependence

$$\mathbf{B} = \varphi([\mathbf{H}_\Sigma]). \quad (1.2)$$

Here φ is a vector function, describing the anisotropic and nonlinear qualities of the cores, square brackets point out that the function is multivalued due to hysteresis.

The EMF created in the sense coil, can be derived from the electromagnetic induction law:

$$u(t) = -w_2 \frac{d\Phi}{dt} = -w_2 \frac{\mathbf{s} \cdot d\mathbf{B}}{dt}, \quad (1.3a) \quad (1.3)$$

where w_2 is the number of turns of the sense coil, $\Phi = \mathbf{B} \cdot \mathbf{s}$ is the magnetic flux in the coils; **s** is the cross-sectional area of the fluxgate cores.

If the anisotropy and hysteresis processes inside the cores are disregarded, instead of the complicated dependence (1.2) a simple equation, concerning only the nonlinear character of the core magnetizing contour is obtained:

$$B = f(H_\Sigma). \quad (1.4)$$

The instantaneous modulus of the field can be found from the expression (1.1):

$$H_\Sigma = \sqrt{H_0^2 + 2H_0 H_1 \cos\alpha + H_1^2}, \quad (1.5)$$

where α is the angle between the vectors $\mathbf{H}_0$ and $\mathbf{H}_1$.

Now let us consider two particular cases of (1.5):

$$\alpha = 0° \Rightarrow H_\Sigma = H_0 \pm H_1 \quad (1.6a)$$
$$\alpha = 90° \Rightarrow H_\Sigma = (H_0^2 + H_1^2)^{\frac{1}{2}}$$

(1.6b)

The first one is characteristic for the fluxgate with parallel fields, while the second one is characteristic for the fluxgate with perpendicular fields.

1.2.1 The Fluxgate with Parallel Fields

First of all, let us consider the fluxgate with parallel fields. For this purpose let us take the most widely used fluxgate with two rod cores (see Figs. 1.1b, 1.3 and 1.5). Consider that the $\mathbf{H}_0$ vector is directed along the longitudinal axis of the cores, and the windings enveloping them have identical axes.

Magnetic induction in each of the cores according to (1.4) and (1.6a) will be

$$\left.\begin{array}{rl} B' &= f(H_0 + H_1), \\ B'' &= f(H_0 - H_1). \end{array}\right\} \quad (1.7)$$

The EMF generated in the sense coil of the fluxgate will be

$$u(t) = -w_2 s \frac{d}{dt}(B' + B'') \qquad (1.3b) \qquad (1.3)$$

where s is the cross-sectional area of one of the cores.

Let us show, that if $H_0 = const \neq 0$ (the external field), the EMF $u(t)$ appears only for a nonlinear dependence $B(H)$.

First, suppose the opposite, i.e. let us consider this dependence be linear: $B = aH$, where a is a constant coefficient. Then

$$B' + B'' = 2aH_0$$

$$u(t) = -2asw \left. \frac{dH_0}{dt} \right|_{H_0=const} = 0$$

Now let us approximate the dependence $B(H)$ by a polynomial of third order [51]:

$$B = aH - bH^3, \qquad (1.8)$$

where a and b are positive approximation coefficients.[4] Then, taking into account the expression (1.7):

$$\left. \begin{array}{rcl} B' &=& aH_0 + aH_1 - bH_0^3 - 3bH_0^2 H_1 - 3bH_0 H_1^2 - bH_1^3, \\ B'' &=& aH_0 - aH_1 - bH_0^3 + 3bH_0^2 H_1 - 3bH_0 H_1^2 + bH_1^3, \end{array} \right\} \qquad (1.9)$$

$$B' + B'' = 2aH_0 - 2bH_0^3 - \underline{6bH_0 H_1^2}. \qquad (1.10)$$

The underlined component in (1.10) is characterized by the product of the external and the drive magnetic fields, and it is this component, which is responsible for the EMF creation in the sense coil:

$$\left. u(t) \right|_{H_0=const\neq 0} = 6bsw_2 H_0 \frac{d}{dt} H_1^{\,2}(t) \neq 0. \qquad (1.11)$$

Evidently, the nonlinear character of the dependence $B(H)$ is really a key factor, responsible for the EMF, containing information about the external field.

1.2.2 The Fluxgate with Perpendicular Fields

Let us consider the fluxgate with perpendicular fields. For this purpose let us use a tube core fluxgate (see Fig. 1.4b).

Let us consider that the $\mathbf{H}_0$ vector is directed along the tube axis, the tube core is homogeneous and the sense coils are strictly perpendicular to the drive coils. On

---

[4]Characterized by simplicity, this approximation is suitable only for qualitative description of processes and phenomena in fluxgates.

1.2. THE FLUXGATE AS A MAGNETIC MODULATOR

fulfilling these conditions and in absence of an external field ($H_0 = 0$), the output EMF will be zero, because the longitudinal component of magnetic induction is zero.

In the presence of an external field ($H_0 \neq 0$) the magnetic induction in the core, according to expressions (1.4) and (1.6b) will be

$$B = f(H_0^2 + H_1^2)^{\frac{1}{2}} . \qquad (1.12)$$

Approximating the dependence $B(H)$ by the polynomial (1.8),

$$B = a(H_0^2 + H_1^2)^{\frac{1}{2}} - b(H_0^2 + H_1^2)^{\frac{3}{2}} .$$ Considering the core isotropy

$$\frac{B_\parallel}{H_\parallel} = \frac{B_\perp}{H_\perp} = \frac{B}{H}$$

where $\parallel$ and $\perp$ are indices of the longitudinal and diametric field induction strength, we find

$$B_\parallel = H_0 \frac{B}{H} = aH_0 - bH_0^3 - \underline{bH_0 H_1^2} . \qquad (1.13)$$

The underlined component in (1.13) contains the product of the external and driving magnetic fields, it can be compared with a similar component in expression (1.10).

The EMF, created in the sense coil of the fluxgate, will be

$$u(t)\Big|_{H_0 = const \neq 0} = bsw_2 H_0 \frac{d}{dt} H_1^2(t) \neq 0 . \qquad (1.14)$$

So, for the fluxgate with perpendicular fields, the nonlinear character of the $B(H)$ dependence will also be a key factor responsible for the EMF containing information about the external field.

1.2.3 Magnetic Modulation

The expressions (1.11) and (1.14), resulting from the approximation (1.8) differ from one another only in the numerical multiple and force a number of general conclusions to be made.

It is evident, that in case of equal values of the measured fields H_0, that the output EMF is proportional to the approximation coefficient b, characterizing the magnetic qualities of the cores, the cross-sectional area of the core, s, w_2 the number of turns of the sense coil and the first derivative from the auxiliary field strength square $H_1(t)$.

In general, the drive field may be given as the sum of two independent sources fields:

$$H_1(t) = H_{m,1} \sin(\omega_1 t) + H_{m,2} \cos(\omega_2 t) , \qquad (1.15)$$

where $H_{m,1}$, $H_{m,2}$ and ω_1, ω_2 are amplitude and frequency accordingly.

Here we shall stick to considering a special case, when $\omega_2 = 0$. Then, a constant (but not the external!) field H_2 is superposed by an alternating field with the amplitude H_m:

$$H_1(t) = H_m \sin(\omega t) + H_2 \,. \tag{1.16}$$

Substituting (1.16) into (1.11) for the fluxgate with parallel fields (when $H_0 = const \neq 0$) then

$$u(t) = 6\omega b s w_2 H_0 \left[2H_2 H_m \cos(\omega t) + H_m^2 \sin(2\omega t)\right] \,. \tag{1.17}$$

A similar expression, but with a different numeric coefficient, can also be made for a fluxgate with perpendicular fields, if the expression (1.16) is substituted into (1.14).

Following equation (1.17) one concludes that the output EMF, of the fluxgate is proportional to the frequency ω of the drive field. Therefore, all other conditions being equal, it is always possible to increase sensitivity by increasing the drive frequency.

It is also clearly seen that the output EMF has a frequency divisible by the drive frequency. This means that in the transformation process, modulation occurs: the zero frequency signal of the external field is transformed to a higher frequency, in this case to the second harmonic of the drive frequency.

Because of their high sensitivity and their modulation properties fluxgates can be easily connected to AC amplifiers. For this reason, it is possible to measure very small values of a constant or slowly changing magnetic field restricted only by the internal magnetic noise of a fluxgate (see Chap. 3).[5]

1.2.4 Two Basic Operation Modes

As derived from expression (1.17) fulfilling the condition (1.16) both the first and the second harmonics of the output EMF can contain information about the external field H_0. It is useful in this connection to consider two different operating modes of a fluxgate [3].

Operating Mode One

The field strength H_m is chosen as small, and H_2 as large, fulfilling condition $H_m \ll H_2$, which makes it possible to disregard the second component of expression (1.17). As a result

$$u_1(t) = 12\omega b s w_2 H_0 H_2 H_m \cos(\omega t) \,. \tag{1.18}$$

---

[5]The fact that fluxgates are magnetic modulators, is reflected in terminology as well: they are sometimes called "ferromodulating transformers". This term, though, can be applied both to fluxgates and magnetic boosters, so we stick to the term "fluxgate" as umambiguous.

1.2. THE FLUXGATE AS A MAGNETIC MODULATOR

Operating Mode Two

The field strength H_m is chosen to be larger than H_0 (a more accurate criterion is given in Chap. 2), and H_2 is equal to zero, ie. $H_m \gg H_0$ and $H_2 = 0$. This results in

$$u_2(t) = 6\omega b s w_2 H_0 H_m^2 \sin(2\omega t) . \tag{1.19}$$

Let us show that in measuring constant or slowly changing magnetic fields the second operating mode of a fluxgate has definite advantages over the first.

Expressions (1.17) and (1.11) assume that the fluxgate cores and windings are exactly the same but, in practice, such an identity cannot be achieved. This leads to the appearance of the false EMF, which is called the misbalance EMF.

Let us find the total EMF (the signal EMF and the misbalance EMF) spectrum, taking into account only the similarity of the cores for the sake of simplicity.

According to expression (1.18), let us introduce approximation coefficients for each core, a' and a'', b' and b'' accordingly. Then, using $a' + a'' = 2a$, $a'' - a' = \varepsilon_a$, $b' + b'' = 2b$, $b'' - b' = \varepsilon_b$, and taking into consideration (1.3), (1.8) and (1.16) let us find the total EMF generated in the sense coil [3]:

$$u_\Sigma(t) = \omega s w_2 \Bigg\{ 6bH_0 \Big[2H_2 H_m \cos(\omega t) + H_m^2 \sin(2\omega t) \Big]$$
$$+ \Bigg[\bigg(\varepsilon_a H_m - \varepsilon_b H_0^2 H_m - \frac{3}{4}\varepsilon_b H_m^3 - 3\varepsilon_b H_2^2 H_m \bigg) \cos(\omega t)$$
$$- 3\varepsilon_b H_2 H_m^2 \sin(2\omega t) + \frac{3}{4}\varepsilon_b H_m^3 \cos(3\omega t) \Bigg] \Bigg\} \tag{1.20a}$$

Let us consider the first and the second operation modes independently. Disregarding all terms containing powers of H_m we derive for the first operating mode ($H_m \ll H_2$):

$$u_\Sigma(t) = \omega s w_2 \Bigg[12bH_0 H_2 H_m \cos(\omega t)$$
$$+ (\varepsilon_a H_m - 3\varepsilon_b H_0^2 H_m - 3\varepsilon_b H_2^2 H_m) \cos(\omega t) \Bigg] \tag{1.20b}$$

For the first operating mode the useful EMF (the first component of the expression) and the misbalance EMF (the second component) have the same frequency ω. It is almost impossible to separate one EMF from the other, and only by decreasing the coefficients ε_a and ε_b, which cannot equal zero, because they depend on lots of internal (the core manufacture) and external factors, it is possible to achieve a high signal-to-noise ratio.

For the second operating mode ($H_\mathrm{m} \gg H_0$, $H_2 = 0$)

$$u_\Sigma(t) = \omega s w_2 \left[6 b H_0 H_\mathrm{m}^2 \sin(2\omega t) \right.$$
$$\left. + \left(\varepsilon_\mathrm{a} H_\mathrm{m} - 3\varepsilon_\mathrm{b} H_0^2 H_\mathrm{m} - \frac{3}{4}\varepsilon_\mathrm{b} H_\mathrm{m}^2 \right) \cos(\omega t) + \frac{3}{4}\varepsilon_\mathrm{b} H_\mathrm{m}^3 \cos(3\omega t) \right]$$
(1.20c)

It is evident, that the useful EMF (the first component of the expression) has a frequency of 2ω, whereas the misbalance EMF s (the second and the third components of the expression) have frequencies of ω and 3ω. Thus, the useful EMF can be easily separated from the misbalance EMF by using a bandpass filter. As a result, the improvement of the signal-to-noise ratio is not restricted to minimizing the coefficients ε_a and ε_b any more. This is the main advantage of the second operating mode.

Expressions (1.20a) and (1.20b) apply to fluxgates with parallel fields. But similar expressions can be easily derived for fluxgates with perpendicular fields. The coefficients ε_a and ε_b express, for instance, the degree of core involution. In the second operating mode these factors appear to be insignificant and do not diminish the high signal-to-noise ratio.

It can be seen that for measuring a constant or slowly changing magnetic field the second operating mode is better. For the first operating mode we can say, that it is quite good for measuring higher frequency fields starting from a few Hz. In this case it has some advantages over the second operating mode. These are: low power consumption [3, §3] and potentially lower levels of internal noise [26], see also Sect. 3.7.

1.3 Ferromagnetic Materials Characteristics

Fluxgate cores are made of ferromagnetic materials iron, nickel, cobalt, gadolinium, their compounds and alloys, using also some manganese, silver, aluminium, etc. alloys are referred to as ferromagnetic substances. The characteristic feature of a ferromagnetic substance is its abnormally high magnetic sensitivity and permeability, reaching values of 10^4–10^5, and sometimes even more. This is why even in weak fields these substances become strongly magnetized and, therefore, gain a strong magnetic induction, reaching a high magnetic energy level. This characteristic is widely used in energy technology, communication and, of course, in measuring technique, magnetic amplifiers and fluxgates.

A typical characteristic of a ferromagnetic material is the closed hysteresis loop, which relates B and H as shown in Fig. 1.6. This family of contours results from the remagnetizing of a substance by an alternating field with discretely increasing amplitude. Inner loops or cycles are called specific. They are not restored after a

1.3. FERROMAGNETIC MATERIALS CHARACTERISTICS

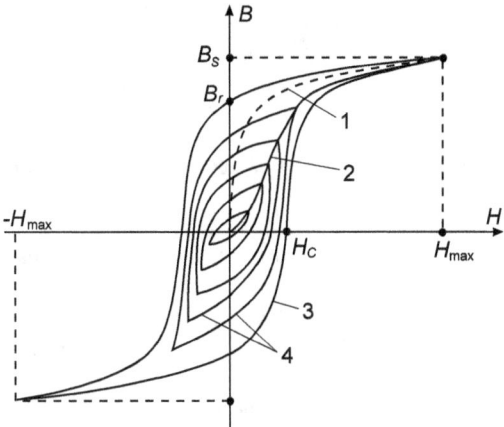

Figure 1.6: The hysteresis curves and the ferromagnetic substance magnetizing curves: 1 – the mean curve; 2 – the main curve; 3 – the maximum loop; 4 – specific loops.

superposition of a constant field for a short period. This behaviour illustrates the multivalued character of the function $B(H)$.

The outer loop, resulting from the $H_m = H_{max}$ drive field, which corresponds to the $\pm B_S$ value on the ordinate axis, is called the *maximum hysteresis loop*. Its characteristic feature is its full restoration after a short-time superposition of a constant field.

The restoration of the maximum loop means that in relation to the alternating field the $B(H)$ function turns to be doublevalued, and $B_2(H_0)$ singlevalued. (B_2 is the amplitude of the second harmonic, causing the output EMF's second harmonic; see Sect. 1.2.) It is the singlevalued nature of this function, which is used in the second operating mode of the fluxgate. Consequently, the second operating mode is characterized, above all, by its ability to provide the singlevalued nature of the transformation, i.e. a high zero stability.

The H_C and B_R values on the maximum loop are called the coercive force and the residual induction (or remanence) of the substance, respectively.

H_C is a measure of the magnetic hardness: materials with $H_C > 10^3$ A/m are considered as magnetic hard whereas materials with $H_C < 10^3$ A/m as magnetic soft. Ferronickel alloys, Permalloys, which are most widely used for core production, are considered to be maximum magnetic soft ($H_C \leq 1$ A/m). Of course high cobalt amorphous alloys can also be used.

B_R or, more precisely, the correlation B_R/B_S, is used to measure the quality of the rectangular loops, showing the anisotropic qualities of a substance or material. For isotropic materials the correlation B_R/B_S is close to 1/2.

The product $H_C \cdot B_S$, proportional to the maximum loop area, serves as a measure of losses on hysteresis. Magnetic soft materials with low values of the coercive force and saturation induction B_S have the minimal losses.

1.3.1 From Hysteresis Loops to Mean Magnetization Curves

The multivalued $B(H)$ dependence, determined by hysteresis, makes functional analysis of ferromagnetic cores more complicated. For this reason hysteresis is sometimes disregarded and a singlevalued $B(H)$ dependence is used.

The curve passing through the tips of the specific cycles (curve 2 in Fig. 1.6) can be accepted as such. This is why this curve is called the main magnetizing curve. It does not differ greatly from the so-called initial magnetizing curve, which results from a smooth field increase on a previously demagnetized sample. The main magnetizing curve can be easily reproduced and, therefore, it is used as a basic characteristic of ferromagnetic materials. It is used in calculating saturated chokes, magnetic screens and similar devices, characterized by a stable working point in relation to constant magnetizing fields. However, where the substance is exposed to cyclic remagnetizing along the maximum hysteresis loop, the main contour of magnetizing leads to discrepancy between calculated and experimental data. In this case, considerably better results are achieved when the mean magnetizing curve is used (curve 1 in Fig. 1.6). The following expression from [36] serves as its definition:

$$B(H) = \frac{B_\uparrow(H) + B_\downarrow(H)}{2}, \qquad (1.21)$$

where the arrows mark the induction values as rising ($\uparrow$) or falling ($\downarrow$) branches of the maximum loop. To simplify calculations the mean magnetizing curve is usually used as an approximation for the maximum loop in functional analysis of fluxgates operating in the second mode.

Since the mean magnetizing curve is an odd function, it can be approximated by an odd powers polynomial:

$$B = \sum_{i=1}^{\infty} a_{2i-1} H^{2i-1}, \qquad (1.22)$$

where a_{2i-1} are approximation coefficients. By using the first three polynomial coefficients a good approximation is obtained. A special case of (1.22) is the shortened polynomial like (1.8), in which $a = a_1$ and $b = -a_3$.

Piecewise linear, arc tangent and other approximations of the average magnetizing contour of the fluxgate cores are also widely used (see Chap. 2).

The choice of approximation of the average magnetizing contour of the cores is determined by the purpose of the calculation.

1.3.2 Normal and Differential Permeabilities of Matter

In practice it is convenient to use not only the B and H values, but also parameters characterizing the magnetic condition of ferromagnetic substances such as the magnetic susceptibility $\hat{\kappa}$ and permeability $\hat{\mu}$, which were mentioned above.

At any point of the mean magnetizing curve of a substance its magnetic condition is characterized by two parameters (disregarding the magnetic anisotropy of the substance):

$$\mu = \frac{1}{\mu_0}\frac{B}{H} \quad \text{and} \quad \mu_d = \frac{1}{\mu_0}\frac{dB}{dH}, \quad (1.23)$$

which are called the normal and the differential permeabilities accordingly. Here μ_0 is a magnetic constant (see Sect. 2.1).

The relation between the permeabilities μ and μ_d is given by the expression

$$\mu_d = \frac{1}{\mu_0}\frac{dB}{dH} = \frac{d}{dH}(\mu H) = \mu + H\frac{d\mu}{dH}, \quad (1.24)$$

which shows that $\mu_d = \mu$ only when $H = 0$ or $d\mu/dH = 0$. In all other cases $\mu_d \neq \mu$, which displays the nonlinear character of the $B(H)$ dependence and shows the necessity for introducing and defining the parameters μ and μ_d.

In Fig. 1.7 you can see the positive branches of the main and the mean magnetizing curves of a substance and the corresponding μ and μ_d permeability dependence on the field strength. The geometrical values of μ and μ_d are defined through the tangents of the angles α and β:

$$\mu = \tfrac{1}{\mu_0}\tan\alpha \quad \text{and} \quad \mu_d = \tfrac{1}{\mu_0}\tan\beta.$$

The angle α is connected with the secant position, going through the coordinate system origin and a chosen point on the $B(H)$ contour, whereas the angle β is connected with the tangent position, going through the same point on the $B(H)$ contour.

The characteristic main magnetizing curve, having two bends (joints), defines the $\mu(H)$ and $\mu_d(H)$ curves – see Figs. 1.7a and 1.7c. Using these curves the following values can be derived:

- the initial magnetic permeability

$$\lim\nolimits_{H \to 0} \tfrac{B}{H}(\tfrac{1}{\mu_0}) = \mu_{\text{init.}} \ ;$$

- the maximum normal permeability

$$\tan\alpha_{\max}(\tfrac{1}{\mu_0}) = \mu_{\max} \ ;$$

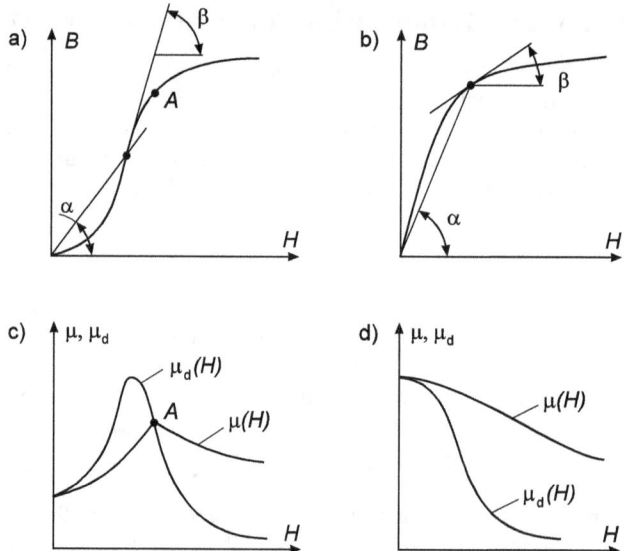

Figure 1.7: The magnetizing curves (a, b) and the substance permeability curve (c, d).

- the maximum differential permeability

$$\tan \beta_{\max}(\tfrac{1}{\mu_0}) = \mu_{d,\,\max}.$$

As can be seen in the figure, $\mu_{d,\,\max}$ is always larger than $\mu_{\max}$ and in a weaker H field. The $\mu(H)$ and $\mu_d(H)$ curves intersect in point A, where $\mu = \mu_{\max} = \mu_d$, because at this point $d\mu/dH = 0$.

The mean magnetizing curve, having one bend (joint), also determines the route of the $\mu(H)$ and $\mu_d(H)$ curves (see Fig. 1.7b and 1.7d). It is characteristic of these curves, that the three permeabilities mentioned above become equal in the initial point $H = 0$:

$\mu_{\text{init.}} = \mu_{\max} = \mu_{d,\,\max}.$

These permeabilities coincide with the $\mu_{d,\,\max} \uparrow$ and the $\mu_{d,\,\max} \downarrow$ permeabilities on the maximum loop of the hysteresis, which, of course, results from the definition of the mean magnetizing curve given in Sect. 1.3.1. The maximum loop stability mentioned there (its independence of short period magnetic fields superposition) determines its presence in the mean magnetizing curve as well, and, therefore, all current values.

From the comparison of the $\mu(H)$ and $\mu_d(H)$ contours, given in Fig. 1.7c and 1.7d it is evident, that these dependencies are characterized by the nonlinear character: approaching one another in weak fields, they separate considerably in strong ones, so, during saturation, μ_d appears to be hundreds of times smaller than μ.

1.4 Demagnetization Factor. Permeability of Cores.

1.4.1 Tensor of a Body's Permeability

Being bodies of finite dimensions, cores of fluxgates are not magnetized in an external homogeneous field in the same way as a material is: the demagnetization factor influences the process of the body's magnetization.

According to V. K. Arkadjev [1], for example, the following relations are valid:

$$\mathbf{B} = \mu_0(\mathbf{h} + \mathbf{J}) , \tag{1.25}$$

$$\mathbf{h} = \mathbf{H} - \hat{N} \cdot \mathbf{J} , \tag{1.26}$$

where $\mathbf{J}$ is the magnetization of the body; $\mathbf{H} = \mathbf{B}_0/\mu_0$ is the strength of the external field; $\mathbf{h}$ is the strength of the internal field (according to Arkadjev "the direct stimulus of magnetism in a substance" [1, §61]); $\hat{N}$ is the tensor of a demagnetization factor connected mainly with the dimensions of a core.

The magnetic susceptibility $\hat{\kappa}$ is defined by

$$\mathbf{J} = \hat{\kappa} \cdot \mathbf{h} = (\hat{\mu} - 1) \cdot \mathbf{h} \quad \text{and} \quad \mathbf{B} = \mu_0 \hat{\mu} \cdot \mathbf{h} , \tag{1.27}$$

with this we get:

$$\mathbf{h} = \left[1 + \hat{N}(\hat{\mu} - 1)\right]^{-1} \cdot \mathbf{H} \tag{1.28}$$

$$\mathbf{B} = \mu_0 \hat{\mu}^* \cdot \mathbf{H} , \tag{1.29}$$

where

$$\hat{\mu}^* = (1 + \kappa)\left[1 + N(\mu - 1)\right]^{-1} \tag{1.30}$$

is the tensor called the magnetic permeability of a body [1].

The reason for introducing the parameter $\hat{\mu}^*$ is to connect the magnetic induction in a body, $\mathbf{B}$, not with hypothetical vector $\mathbf{h}$ as in (1.27), but with the real vector $\mathbf{H}$ which has a dimension.

Considering (1.30) it is evident that $\hat{\mu}^*$ takes account not only of the magnetic anisotropy of the substance but also of the anisotropy of the body's shape, ie. of its different dimensions in different directions.

In some cases (even when applied to materials used in fluxgates) the anisotropy of the substance can be neglected. Then $\hat{\mu} = \mu$ and $\hat{\mu}^*$ takes account only of the anisotropy of the shape

$$\hat{\mu}^* = \frac{\mu}{1 + \hat{N}(\mu - 1)} \, . \tag{1.31}$$

Referred to the axes (x, y, z) of the body, this tensor has the followings components:

$$\begin{aligned} \mu_x^* &= \frac{\mu}{1+N_x(\mu-1)} \\ \mu_y^* &= \frac{\mu}{1+N_y(\mu-1)} \\ \mu_z^* &= \frac{\mu}{1+N_z(\mu-1)} \end{aligned} \tag{1.32}$$

where N_x, N_y and N_z are the components of the tensor, the so-called coefficients of demagnetization. As a result, if a body is isotropic (spheric, for example) then $N_x = N_y = N_z$ and $\mu_x^* = \mu_y^* = \mu_z^*$; if a body is anisotropic (for example, if it has the shape of a plate – a right-angled parallelepiped with unequal sides), then $N_x \neq N_y \neq N_z$, and, consequently, $\mu_x^* \neq \mu_y^* \neq \mu_z^*$.

According to (1.29)

$$B_x = \mu_0 \mu_x^* H_x \, , \qquad B_y = \mu_0 \mu_y^* H_y \, , \qquad B_z = \mu_0 \mu_z^* H_z \, , \tag{1.33}$$

if $\mu_x^* \neq \mu_y^* \neq \mu_z^*$, then $\mathbf{B} \not\parallel \mathbf{H}$, ie. vector $\mathbf{B}$ is not parallel to $\mathbf{H}$ (the vector of the external field).

When the anisotropy of a body is large, for example, in a long thin rod orientated in the direction of one axis, components B_y and B_z may be neglected and it is possible to assume that the body is magnetized by the appropriate projection of the vector

$$B = \mu_0 \mu_x^* H_x = \mu_0 \mu^* \mathbf{H} \cdot \mathbf{x}^0 \, . \tag{1.34}$$

We shall return to this question in Sect. 2.7 when the directional characteristics of fluxgates are considered.

1.4.2 Normal and Differential Permeabilities of a Body

The characteristics of a ferromagnetic body can be defined in the same way as the characteristics of a ferromagnetic substance were defined (see Sect. 1.3). In Fig. 1.8a there are the medium curves of the magnetization of a substance and a body constructed according to ratios (1.27)–(1.29), where $\hat{\mu} \equiv \mu$ and one of the axes of symmetry of the body is superposed with vector $\mathbf{H}$.

It can be seen that, to achieve a certain magnetic induction, a higher field strength must be applied to a body than to a substance. According to (1.28) the ratio $1/[1 + N(\mu - 1)]$ serves as a link between h and H. This is why the curve of magnetization of a body is none other than the curve of magnetization of a

1.4. DEMAGNETIZATION FACTOR. PERMEABILITY OF CORES.

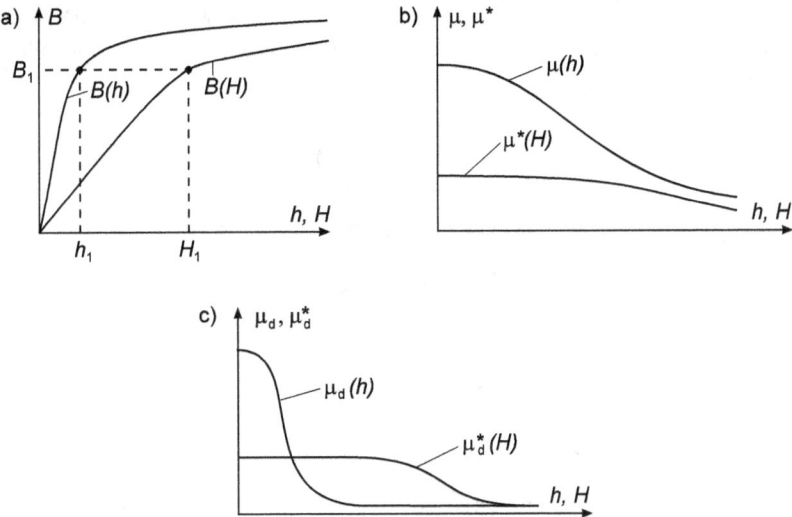

Figure 1.8: Magnetization curves (a) and permeability curves of a substance and a body (b, c).

substance shifted to the right. Only when $N = 0$ (an extremely long rod) the contours will overlap.

In Fig. 1.8b dependencies $\mu(h)$ and $\mu^*(H)$ suitable for contours $B(h)$ and $B(H)$ are shown; in Fig. 1.8c – dependencies $\mu_d(h)$ and $\mu_d^*(H)$.

Parameters μ^* and μ_d^* are the analogous of parameters μ and μ_d introduced in Sect. 1.3.2:

$$\mu^* = \frac{1}{\mu_0}\frac{B}{H} \quad \text{and} \quad \mu_d^* = \frac{1}{\mu_0}\frac{dB}{dH}. \qquad (1.35)$$

Correspondingly they are called the normal and differential permeabilities of a body. Parameter μ^* is defined with the help of ratio (1.31) and parameter μ_d^* with the help of then differential analogue of (1.31):

$$\mu_d^* = \frac{\mu_d}{1 + N(\mu_d - 1)} \qquad (1.36)$$

One can see that the righthand side of (1.35) coincides with the corresponding side of (1.23). The latter is a specific case of (1.35) which is true if $N = 0$. As applied to bodies of finite dimensions ($N \neq 0$) parameters μ and μ_d should be defined through the B/h and dB/dh ratios.

According to (1.24) we have

$$\mu_d^* = \mu^* + H\frac{d\mu^*}{dH}. \qquad (1.37)$$

In Fig. 1.9 there is a family of curves $\mu^*(\mu)$ constructed according to (1.31). In view of the identical structure of (1.31) and (1.36) an analogous family can be built for the dependence $\mu_d^*(\mu_d)$. Through the course of the curves one can see that with

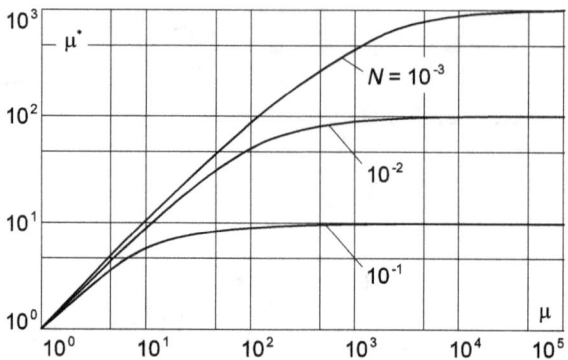

Figure 1.9: Curves which characterize the dependence of the body permeability on the substance permeability for different values of the demagnetization coefficient N.

increasing substance permeability μ, the permeability of body μ^* approaches some ultimate value

$$\lim_{\mu \to \infty} \mu^* = \frac{1}{N} = m \qquad (1.38)$$

(parameter m is called the permeability of the shape[6]).

For attaining the maximum magnetic induction in a body the conditions (1.38) are targeted. However, a better sense of how to maximize magnetic induction exists.

From (1.31) one can see that $\mu^* = f(\mu, N)$, therefore

$$\Delta \mu^* = \frac{\partial f}{\partial \mu} \Delta \mu + \frac{\partial f}{\partial N} \Delta N , \qquad (1.39)$$

where

$$\frac{\partial f}{\partial \mu} = \frac{1 - N}{[1 + N(\mu - 1)]^2} \quad \text{and} \quad \frac{\partial f}{\partial N} = \frac{\mu(\mu - 1)}{[1 + N(\mu - 1)]^2} . \qquad (1.40)$$

If $N \leq 10^{-2}$ ($m \geq 10^2$) and $\mu \geq 10^4$ (the characteristic values for the cores of fluxgates) we have

$$\frac{\Delta \mu^*}{\mu^*} = \frac{1}{\mu N} \frac{\Delta \mu}{\mu} - \frac{\Delta N}{N} . \qquad (1.41)$$

---

[6]Parameter m has a dimension inverse to N and should be called the "receptivity of the form" (according to the definition given in [1, p. 115, 121]). But in the SI system of units there is no difference between the values κ^* and μ^*: they are equal to m. This is why parameter m can be called "permeability of the shape"

1.4. DEMAGNETIZATION FACTOR. PERMEABILITY OF CORES.

According to (1.41) the relative increment of the body permeability is μN times weaker than the relative increment of the corresponding substance permeability, and is proportional to the relative increment of the body's demagnetization coefficient.

Thus, the curve displacement of ferromagnetic magnetization, appearing if $N \neq 0$, leads to the stabilization of body permeability and it is used when fluxgates with a stable coefficient of transformation are required.

1.4.3 Autostabilization of Body Permeability

According to (1.26) a body, being magnetized in an external field of strength H, creates its own demagnetizing field $H_{\text{demag}} = NJ$ aligned inside the body with the external field. Consequently, the coefficient of demagnetization N can be defined as ratio H_{demag}/J, with the permeability of form m as the inverse ratio J/H_{demag} (see Fig. 1.10a). As the strength of the internal field is defined by ratio $h = H - NJ$,

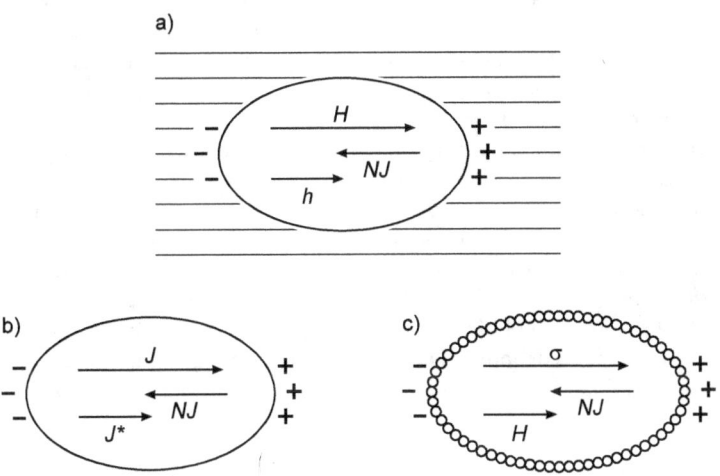

Figure 1.10: The action of magnetic feedback on a body which is magnetized by the external field (a) and has residual magnetization (b), and on a coil (c).

h decreases with an increase in N. The cancellation of the body length in the considered direction allows to consider the mentioned dependence as the negative magnetic feedback. This feedback is determined by the product μN as derived in (1.41).

In long thin bodies the feedback in a longitudinal direction is weaker than in transversal directions. This leads to the previously considered anisotropy, when one has to take account of not one but at least three parameters of the body μ_x^*, μ_y^* and μ_z^* [ratio (1.32)].

With the help of negative feedback it is possible to explain the stabilization effect of body permeability. According to (1.41) if $\mu N \gg 1$ the negative feedback becomes so strong that the changes of substance permeability (for example, under the influence of temperature) lead to such slight changes of body permeability that they can be practically neglected.

The "coefficient of demagnetization" and "magnetic negative feedback" terms are spread to other cases as well.

In Fig. 1.10b there is a body which possesses residual magnetization (a permanent magnet). If $H = 0$ the action of the inverse connection leads to the fact that the resultant magnetization or the magnetization of the body is:

$$J^* = (1 - N)J .\tag{1.42}$$

This expression can be derived if we substitute (1.26) for (1.25). From (1.42) it is evident that with the permanent residual magnetization of the substance, J, the more the coefficient of demagnetization N the less the magnetization of body J^* is. But, the greater the value of N the less is the value of μ_d [according to (1.36)]. Consequently, ΔB increments caused by the influence of external fields to the residual induction of the body, B_R^*. It is necessary to notice that coefficient N is defined by ratio H_{demag}/J as before.

Magnetization J is directly connected with microscopic currents flowing inside a substance (Sect. 2.1). These currents can be correlated with surface currents which flow around the body. If the body is magnetized homogeneously the magnetization J is numerically equal to the uniform density of surface current $\sigma = I_\Sigma/l$, where I_Σ is total surface current and l is the length of the body in the direction of magnetization.

In Fig. 1.10b there is a bobbin with current equivalent (according to Ampere's principle) to a magnetized body. The strength of the field inside a coil can be found through the following equation

$$H = (1 - N)\sigma ,\tag{1.43}$$

where $\sigma = Iw/l$; w is the number of coil turns. From this expression one can see that $\sigma = H$ only if $N = 0$, ie. for an extremely long or closed (for example, toroidal) coil (V. K. Arkadjev called $\sigma = H$ "normal" [1, §29]). In all other cases (except $N = 1$) $H < \sigma$, which also indicates the action of the negative feedback. Coefficient N is defined by ratio $(\sigma - H)/\sigma$.

Thus, coefficient N, ie. $\hat{N}$, serves first of all as a geometric factor which allows to take into account the influence of the body's shape on its magnetic characteristics. This factor can be applied to bodies magnetized in an external field, to previously magnetized bodies (magnets) and to coils with currents. In bodies of a substance with magnetic permeability $\mu > 1$, and in coils with current (if $\mu = 1$) the presence of the factor ($N \neq 0$) leads to the appearance of a negative field which is coordinated with the conception of magnetic negative feedback.

1.4.4 Calculation of the Permeability of Ellipsoidal Cores

The exact calculation of the demagnetization coefficient N or inverse quantities – the permeabilities of shape $m = 1/N$ – is possible only for bodies which are homogeneously magnetized in an external homogeneous field. Such bodies are tri-axial ellipsoids and ellipsoids of rotation [1, 34].

The following equation

$$N_a + N_b + N_c = \frac{1}{m_a} + \frac{1}{m_b} + \frac{1}{m_c} = 1 \tag{1.44}$$

is true for any ellipsoid with axes a, b and c; one can see from it that the largest ultimate value of each of three coefficients $N_i = 1/m_i$ (with the minimization of other two) is unity.

For the ellipsoid of rotation stretched along the axis a (ovoid) we have [1, §70]:

$$m_a = \frac{\lambda^2 - 1}{\frac{\lambda}{\sqrt{\lambda^2 - 1}} \ln(\lambda^2 + \sqrt{\lambda^2 - 1}) - 1} \tag{1.45a}$$

If $\lambda \geq 10$ we can use formula

$$m_a \approx \frac{\lambda^2}{2\ln\lambda - 1}, \tag{1.45b}$$

instead of expression (1.45), which gives values m_a with an error less than 1%. With regard to (1.44)

$$m_b = m_c = 2\frac{m_a}{m_a - 1}, \tag{1.46}$$

where $\lambda = a/b = a/c$. For a sphere ($a = b = c$), we can get from (1.44):

$$m_a + m_b + m_c = 3. \tag{1.47}$$

For a spheroid, we have [1]:

$$m_a = \frac{1 - \lambda^2}{1 - \frac{\lambda}{\sqrt{1 - \lambda^2}} \arccos\lambda} \tag{1.48a}$$

If $\lambda \leq 0.1$ instead of (1.48a) we have

$$m_a \approx \frac{2}{2 - \pi\lambda}$$
(1.48b)

and according to (1.46) we have

$$m_b = m_c \approx \frac{4}{\pi\lambda} \, .$$
(1.49)

In case of an extremely long disk, for example, we get:
$m_a = 1$ and $m_b = m_c = \infty$.

For a tri-axial ellipsoid stretched in direction ($a \neq b \neq c$) if $a \gg b$ and $a \gg c$ [34] we have

$$m_a \approx \frac{a^2}{bc\left(\ln\dfrac{4a}{b+c} - 1\right)} \, ,$$
(1.50)

if $b = c$ this expression turns into (1.45b).

Finally, for a cylinder with elliptic cross-section stretched along the axis a we have [1, p. 135]

$$m_a \to \infty \, ; \qquad m_b \approx 1 + \lambda_\perp \, ; \qquad m_c \approx 1 + \frac{1}{\lambda_\perp} \, ,$$
(1.51)

where $\lambda_\perp = b/c$. These ratios comply with (1.44). In particular, if $\lambda_\perp = 1$ (a strongly stretched cylinder with circular cross-section) we get $m_b = m_c \approx 2$.

The process of manufacturing ellipsoidal cores is very difficult. This is why non-ellipsoidal cores – long narrow stripes (plates) or stretched cylinders of wire are used in fluxgates. The calculation of the shape permeability of such cores is given below.

1.4.5 Calculation of the Permeability of Non-ellipsoidal Cores

It should be noticed that strictly analytical expressions for the calculation of the form permeability of non-ellipsoidal cores do not exist. Such cores gain non-homogeneous magnetization in an homogeneous external field, and it is therefore possible to introduce different coefficients N for different parts of a core and, consequently, different values m. For this reason two new parameters have to be introduced in application to non-ellipsoidal cores which are: m, the central permeability of shape, and $\overline{m}$, the mean ie. averaged over the core length) permeability of shape [1].

The permeability of non-ellipsoidal cores turns to be dependent not only on the geometric dimensions of the body but also on the permeability μ of the substance

1.4. DEMAGNETIZATION FACTOR. PERMEABILITY OF CORES.

which it is made of. Since a change of μ leads to the redistribution of magnetization J this fact complicates calculations. Only if $\mu \to \infty$ it is possible to speak about definite steady values N_∞ and m_∞ [1, 36]. The condition $\mu \to \infty$ is achieved in fluxgates with cores made from alloys of high permeability (μ has order 10^5). This is why, in future, we shall use only N_∞ and m_∞ omitting indices ∞.

Although strictly analytical expressions for the calculation of the form permeability or coefficient N of non-ellipsoidal cores do not exist there is a relationship between non-ellipsoidal and ellipsoidal cores. An example of such a relationship is given in Fig. 1.11 where a curve of relations between the permeabilities of cylindrical and ellipsoidal forms (ovoid) with the same values $\lambda = l/d = a/b$ where l is the length and d is the diameter of a cylinder, built according to the facts of work [1, §78], is depicted. One can see that values m_{cyl} are always bigger than values m_{ell}, and if λ increases the $m_{\text{cyl}}/m_{\text{ell}}$ ratio decreases. A similar relationship also exists between the permeabilities of laminar, elliptic-cylindrical and tri-axial ellipsoidal cores. These relationships serve as a basis for investigating the $m(\lambda)$ or $m(\lambda')$ semi-empirical dependencies, where $\lambda' = l/\sqrt{s}$; s is the cross-sectional area of a core. These dependencies are suitable for the calculation of the form permeability of non-ellipsoidal cores [1, 28, 34]. The semi-emperical formula by M. A. Rozenblat is best [34].

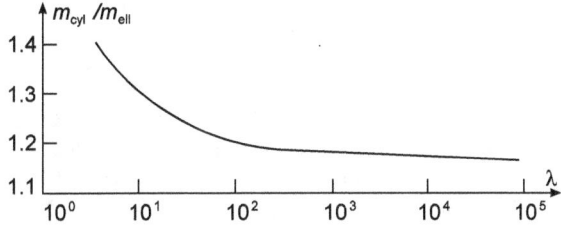

Figure 1.11: Dependence of the ratio of permeabilities, $m_{\text{cyl}}/m_{\text{ell}}$, on the relative length of the body.

Using (1.50) as a prototype and equating the $\lambda' = l/\sqrt{s}$ values of different bodies, Rozenblat found a universal equation for the calculation of the permeability of ellipsoids, solid and hollow cylinders and plates of right-angled cross-sections

$$m = \frac{\pi l^2}{4s \left(\ln \frac{kl}{\delta + \rho} - 1 \right)}, \quad (1.52)$$

where l is the length; s is the cross-sectional area (for ellipsoids – equatorial); δ and ρ are orthogonal dimensions of the core; $k = 4 - 0.732(1 - e^{-5.5\sigma/\rho})$. A comparison of calculated and measured values shows that the definition error of the m permeability according to formula (1.52) does not exceed 3–4 % and it is quite suitable for use.

Cores	ν	k [34] Circular and elliptical cross-section	k [34] Right-angled cross-section
Rod	1.0		
Circular and ellipsoidal (if $a/b \leq 3$)	1.3	2.4	3.6
Frame (square)	1.4		

Table 1.1: Coefficients for the calculation of the form permeability of cores according to (refeqn1.53)

We introduced the coefficient ν in (1.52) to envelope closed cores as well:

$$m = \frac{\nu \pi L^2}{4s \left(\ln \dfrac{kl}{\delta + \rho} - 1 \right)} , \qquad (1.53)$$

where L is the extent of a core in the direction for which m is calculated (for a circular core $L = D$, where D is the diameter); ν is the coefficient (table 1.1). Coefficient ν has, according to indirect measurement, a 7–10 % error. Formula (1.53) differs from others suggested for circular cores [5, 59] not only by its universality but by the fact that the influence of the cross-section of the cores is taken into account. This is important because closed cores are made, as a rule, from a thin Thermalloy tape, so the dimensions δ and ρ differ.

Formulae (1.52) and (1.53) are suitable for calculations when a coil, covering a core, is concentrated in its central part. If a winding covers a larger area remaining symmetrical relative to the core center, one should use the following formula for the middle permeability of the form of non-ellipsoidal cores [34, 20]:

$$\overline{m} = \nu \frac{\pi (L^2 - 0.25 \, l_w^2)}{4s \left(\ln \dfrac{kl}{\delta + \rho} - 1 \right)} , \qquad (1.54)$$

where l_w is the coil length.

Formula (1.54) takes into account the heterogeneity of magnetic induction along the length of non-ellipsoidal cores (Fig. 1.12). As a rule, sense coils occupy a length l_w less than half of the length of core L. The difference between the m

1.4. DEMAGNETIZATION FACTOR. PERMEABILITY OF CORES.

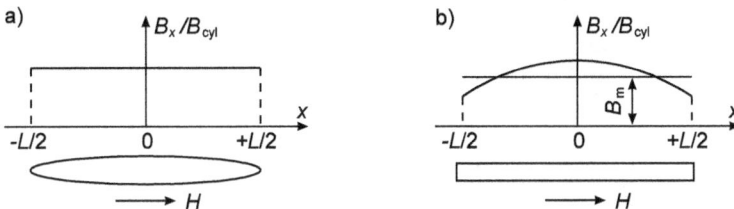

Figure 1.12: The distribution of magnetic induction along ellipsoidal (a) and non-ellipsoidal (b) cores.

and $\overline{m}$ values in this case will not exceed 6 %. If $l_w < 1/4$ it will become less than 1.5 %, so, practically, $\overline{m} \approx m$ and it is possible to use only (1.53).

In practice, the stabilization of the permeability μ^* and μ_d^* is very important. According to Sect. 1.4.2 it is recommended to choose values m within limits 10^2–10^3. Then if $\mu_{\max} = 10^4 \ldots 10^5$ (permalloy, high cobalt amorphous alloys), the ratio $\mu_{\max}/m \geq 10^2$. According to the expressions mentioned in Sect. 1.4.2 with an error not exceeding 1 %, we derive

$$\mu_{\max}^* = \mu_{d,\max}^* \approx m . \tag{1.55}$$

Ratio (1.55) means that the calculation of the body permeability is the same as the calculation of the permeability of the cores shape.

Chapter 2

Fundamentals of Parametric Theory

2.1 Magnetic Quantities and Parameters

The magnetic induction **B** and magnetizing force **H** are fundamental quantities that are used to describe transformations occurring in magnetic media. These physical quantities were introduced into electrodynamics, including Maxwell's equations, independently from each other.

The vector **B** was introduced to represent a quantity, which is both acting within a medium and being dependent upon the medium properties. This quantity may be defined either via the mechanical force $d\mathbf{F}$ applied to the element of current $I\,d\mathbf{l}$ (Ampere's law):

$$d\mathbf{F} = I\,d\mathbf{l} \times \mathbf{B}, \tag{2.1}$$

From this expression one may see that the magnetic induction is measured either in Newtons per Ampere per meter or in Volt-seconds per square meter in SI units. These identical units are called "Tesla" (T).

The vector **H** was introduced to represent a quantity which is directly related to the macroscopic currents (the law of Biot-Savart):

$$d\mathbf{H} = \frac{1}{4\pi} \frac{I\,d\mathbf{l} \times \mathbf{r}^0}{r^2} \tag{2.2}$$

where the value $I\,d\mathbf{l}$ is the current element, r is the distance to the point where the magnetizing force is calculated, $\mathbf{r}^0$ is the unit vector. The total value of the magnetizing force at this point is found by integration of the right side of (2.2) along the contour L with current I ($L = \int dl$). The field magnetizing force **H** does not depend upon the properties of the medium and, according to (2.2), is measured in Amperes per meter.

The introduction of **B** and **H** requires to establish a relationship between them, and consequently a description of the magnetic properties of the medium.

2.1. MAGNETIC QUANTITIES AND PARAMETERS

In many cases, including ferromagnetic media, this relationship has a complicated nature:

- the vectors **B** and **H** turn out not to be parallel (anisotropic media),
- there is no direct linear dependence between B and H (non-linearity), and, finally,
- different values of B may correspond to the same value of H (hysteresis). This relationship between **B** and **H** was taken into account in Chap. 1 in (1.2).

Instead of (1.2) the following equation is frequently used:

$$\mathbf{B} = \mu_0(\mathbf{H} + \mathbf{J}), \qquad (2.3)$$

where the apparent simplification is achieved by virtue of an introduction of yet another vector value, namely the magnetization **J**. The coefficient μ_0, which is of minor importance for now, will be considered later.

The vector **J** is directly related to the microscopic currents (or the molecular currents, according to Ampere's hypothesis), which exist in the medium. Unlike the macroscopic currents, which the vector **H** is related to, the microscopic currents cannot be measured and one has to evaluate them on the basis of the magnetization **J**.

The relationship between magnetization **J** and the magnetizing force **H** is analogous to (1.2):
$\mathbf{J} = \mathbf{\Psi}([\mathbf{H}])$,
where the vector function $\mathbf{\Psi}$ is used to describe the anisotropic and non-linear properties of the medium, the square brackets indicating that $\mathbf{\Psi}$ is a multivalued function due to hysteresis.

Let us ignore the hysteresis, i.e. assume that $\mathbf{\Psi}$ is a single-valued function. Then, the relationship between the quantities of **J** and **H** will be

$$\mathbf{J} = \hat{\kappa}(\mathbf{H}) \cdot \mathbf{H}. \qquad (2.4)$$

Here $\hat{\kappa}$ represents the magnetic susceptibility of the medium with its components being dependent on the components of vector **H**.

Substituting (2.4) into (2.3) a similar relationship can be derived for **B** and **H**

$$\mathbf{B} = \mu_0[\mathbf{H} + \hat{\kappa}(\mathbf{H}) \cdot \mathbf{H}] = \mu_0 \hat{\mu}(\mathbf{H}) \cdot \mathbf{H}, \qquad (2.5)$$

where $\hat{\mu} = \hat{1} + \hat{\kappa}$ is the magnetic permeability of the medium. The tensor components of $\hat{\mu}$ depend on the components of the vector H.

The magnetic susceptibility and the magnetic permeability are parameters of the medium. We already used them in the previous chapter. They are quite similar to the other media parameters, such as the dielectric permeability and the

34 CHAPTER 2. FUNDAMENTALS OF PARAMETRIC THEORY

electrical or thermal conductivity. They are similar to electric circuit parameters, such as resistance, capacitance and inductance.[1]

It is well known that all substances have magnetic properties. In this respect substances are usually divided into weak and strong magnetic substances. Ferromagnetic materials belong to the class of strong magnetic substances. The magnetic permeability of a substance serves as a measure of its "magnetic properties".

The modern science of magnetism is based on a complete rejection of the concept of free magnetic charges, the magnetic phenomena being approached exclusively from the point of view of an interaction among electric currents.

The equation for a force, i.e. electrodynamic or magnetic, between two current elements $I_1\, d\mathbf{l_1}$ and $I_2\, d\mathbf{l_2}$ in vacuum has the following form:

$$d\mathbf{F} = \frac{\mu_0}{4\pi} \frac{I_2\, d\mathbf{l_2} \times (I_1\, d\mathbf{l_1} \times \mathbf{r}^0)}{r^2} , \qquad (2.6)$$

where $\mu_0 = 4\pi 0^{-7}$ H/m is the magnetic constant, r is the distance between the currents, $\mathbf{r}^0$ is a unit vector.

The constant μ_0, which may be referred to as the electrodynamic constant, entered (2.6) as a result of the use of the practical units such as Ampere, Volt, Henry, Farad, etc., which were formerly used with the newly introduced mechanical and magnetic units. Indeed, if one expresses the force $\mathbf{F}$ in Newtons, the current I in Amperes and the distance r in meters, then the constant μ_0 will be expressed in Amperes square (which directly follows from the equation above) or in Henry per meter (as is commonly used in scientific literature).

Equation (2.6) contains both Ampere's and Biot-Savart's law. For this reason it should agree with (2.1) and (2.2). However, due to the difference in the measurement units for $\mathbf{B}$ and $\mathbf{H}$, such an agreement is possible only if the following equality is used

$$B_0 = \mu_0 \mathbf{H} , \qquad (2.7a)$$

which is a special case of (2.3).

It is obvious that the fields created by current I_1 (in accordance with Biot-Savart's law) at a certain point in vacuum space and the field caused by the current I_2 at this point (accordingly to Ampere's law) are the same magnetic fields. Hence,

---

[1] In physics and in technical areas parameters are introduced to describe the quantities characterizing the stable properties of objects. In mathematics by parameters one means the coefficients, which are used in a formula together with basic variables. The two definitions do not contradict to each other and correspond fully to the concepts developed in the present work.

2.1. MAGNETIC QUANTITIES AND PARAMETERS

Biot-Savart's law should be rewritten in the form[2]

$$d\mathbf{H}^* = \frac{\mu_0}{4\pi} \frac{I\, d\mathbf{l} \times \mathbf{r}^0}{r^2} \qquad (2.2b)$$

incorporating the constant μ_0 directly into the body of the expression and assigning the same dimension to both the magnetizing force $\mathbf{H}^*$ and the magnetic induction $\mathbf{B}$.

Such a definition of the magnetizing force provides the following advantages:

1. The magnetizing force would return to its original meaning as a forcible characteristic of the field in vacuum (presently it is an estimated characteristic, as is $\mathbf{B}$, which is acting upon the primary magnetomeasuring converters, but not$\mathbf{H}$, and would allow Biot-Savart's law to be experimental.

2. It would replace (eqn2.7a) with the identity

   $$\mathbf{B}_0 \equiv \mathbf{H}^* \qquad (2.7b)$$
   and use only one magnetic vector, which is vector $\mathbf{H}^*$, excluding $\mathbf{B}_0$ completely.

3. It would permit the exclusion of the formally introduced notion of the "absolute magnetic permeability", as well as the possibility of treating the constant μ_0 as a magnetic parameter of a vacuum.

Unfortunately, the new definition of the magnetizing force is not accepted yet.

In the present book we have to use both $\mathbf{B}_0$ and $\mathbf{H}$ for the description of the magnetic field in vacuum. $\mathbf{B}_0$ is the experimentally observed quantity. It is this value that will be used eventually to describe the characteristics of fluxgates and fluxgate type devices. $\mathbf{H}$ is considered as the value inferred in the process of an experiment, ie. estimated value, which is calculated and represents a convenient tool for calculations.[3]

We also restrain ourselves from the use of the terms "absolute" and "relative magnetic permeability", simply speaking about the magnetic permeability as a measureless parameter of a medium.

---

[2] See, for example, L. A. Sena "Edinitsy fizicheskikh velitchin i ikh razmernost", Moscow, Nauka, 1969, p. 202 (in Russian) and Y. V. Afanassiev, Geofizicheskaya apparatura, 1983, issue 77, p. 128 - 134 (in Russian).

[3] It would be possible to use only one vector $\underline{\mathbf{B}}_0$ for the description of the magnetic field in vacuum. However, in a medium two quantities are principally needed as shown. To use the vectors $\mathbf{B}$ and $\mathbf{B}_0$ for this purpose would be inconvenient due to the coincidence of notations for the function and the argument (the information on the argument is carried about by only the index "0").

2.2 The Influence upon Permeability. The Essence of the Parametric Approach

The influence upon the substance permeability μ, as well as that of the body μ^* of a ferromagnet, may be achieved by applying a magnetic field (see Fig. 1.7, Fig. 1.8), or by other causes, such as, for example, application of mechanical forces (Villary's effect) or providing thermal changes near the Curie's point (Gopkinson's effect) [4, Sect. 4-9], [40, Sect. 9-1].

Fluxgates are converters, where the influence upon the parameter $\hat{\mu}^*$ is achieved by imposing a magnetic field.

Let the cores of the fluxgates be effected by a strong field $H_1(t)$ alternating with a constant amplitude. Then one may write

$$\hat{\mu}^*[H_1(t)] = \hat{\mu}^*(t) . \tag{2.8}$$

Hence, in the correspondence with the expression (1.29),

$$\mathbf{B}(t) = \mu_0 \hat{\mu}^*(t) \cdot \mathbf{H}_0(t) ; \tag{2.9}$$

$$u(t) = -sw\mu_0 \mathbf{i}_w^0 \left(\hat{\mu}^*(t) \frac{d\mathbf{H}_0}{dt} + \mathbf{H}_0(t) \frac{d\hat{\mu}^*}{dt} \right) , \tag{2.10}$$

where $\mathbf{i}_w^0$ is a unit vector, which coincides with the axis of the sense coil of the converter. (s, w as defined in Sect. 1.2.1.)

Expressions (2.9) and (2.10) are linear equations with constant coefficients. They are similar to those equations used to describe processes and phenomena in parametric circuits. Thus, (2.10) is nothing else but a parametric form of the electromagnetic induction law. It is seen from this equation that the EMF $u(t)$ at the fluxgate output appears in cases when $\mathbf{H}_0 = const \neq 0$. In this case the EMF, $u(t)$, is proportional to the derivative $d\,hat\mu^*/dt$, i.e. to the rate of change of the parameter $\hat{\mu}^*$.

2.2.1 Transversal Excitation

The advantages of the second operational mode of the fluxgates were mentioned in Sects. 1.2 and 1.3, when the additional (non-measurable!) constant field $H_2 = 0$ and the amplitude of the alternating field $H_1(t) = H_m \sin(\omega t)$ dominates the measured constant field H_0: $H_m \gg H_0$, .

The second mode of operation is consequently the one in which the alternating field $H_1(t)$ provides the main influence upon $\hat{\mu}^*$. This is why in case of the second operational mode this field is referred to as the excitation or drive field.

First we will consider the equation of conversion for fluxgates with transversal excitation (see Fig. 1.4).

2.2. THE INFLUENCE UPON PERMEABILITY

G. S. Gorelik [15] gave the basics of the theory of such fluxgates. It was shown that, with respect to longitudinal measured field $H_0(t)$, the lateral excitation field $H_1(t)$ influences the parameter μ^*, so for longitudinal magnetic induction we have

$$B(t) = \mu_0 \mu^*[H_1(t)] H_0(t) . \tag{2.11}$$

The expression for the output EMF will be the following:

$$u(t) = -sw_2\mu_0 \left(\mu^*[H_1(t)] \frac{dH_0}{dt} + H_0(t)\frac{d\mu^*}{dt} \right) . \tag{2.12}$$

It follows from here that the EMF $u(t)$ also appears when $H_0 = const \neq 0$ and that it is proportional to dmu^*/dt.

Using the shortened polynomial approximation [see (1.8)] for the mean curve of the core magnetizing $B(H)$, for $H_1(t) = H_m \sin(\omega t)$ we have

$$\mu^*[H_1(t)] = \frac{1}{\mu_0}\frac{B}{H} = \frac{1}{\mu_0}\left(a - \frac{1}{2}bH_m^2 + \frac{1}{2}BH_m^2 \cos(2\omega t) \right) . \tag{2.13}$$

It can be seen that the second harmonic is contained in the periodic function $\mu^*[H_1(t)]$. Due to this, for $H_0 = const \neq 0$, we find from (2.12)

$$u(t) = -sw_2\mu_0 H_0 \frac{d\mu^*}{dt} = \omega b s w_2 H_0 H_m^2 \sin(2\omega t) . \tag{2.14}$$

This expression coincides with (1.14) that was derived by a non-linear approach.

2.2.2 Longitudinal Excitation

The theory of fluxgates with longitudinal excitation was developed by R. I. Yanus, L. Kh. Fridman and V. I. Drozhzhina [3]. Using the expansion of the function $B(H_\Sigma)$, where $H_\Sigma = H_1 + H_0$ into a Taylor series with respect to the small influence H_0 we have

$$B(H_1 + H_0) = f(H_1) + \sum_{n=1}^{\infty} f^n(H_1)\frac{H_0^n}{n!} \tag{2.15}$$

(n is the order of the derivative). By restricting themselves to the first two terms in the expansion and assuming small values of H_0 and two rods with identical cores and windings (see Figs. 1.1b, 1.3 and 1.5), the authors of [3] derived the following expressions

$$B'(H'_\Sigma) = f'(H_1 + H_0)$$
$$= \mu_0\left(\mu^*[H_1(t)]\,H_1(t) + \mu_d^*[H_1(t)]\,H_0(t) - \frac{d\mu_d^*}{dH_1}\frac{H_0^2(t)}{2}\right),$$

$$B''(H''_\Sigma) = f''(-H_1 + H_0)$$
$$= \mu_0\left(-\mu^*[H_1(t)]\,H_1(t) + \mu_d^*[H_1(t)]\,H_0(t) + \frac{d\mu_d^*}{dH_1}\frac{H_0^2(t)}{2}\right),$$

$$B_\Sigma(t) = B'(H'_\Sigma) + B''(H''_\Sigma) = 2\mu_0\mu_d^*[H_1(t)]\,H_0(t), \qquad (2.16)$$

$$u(t) = -2sw_2\mu_0\left(\mu_d^*[H_1(t)]\frac{dH_0}{dt} + H_0(t)\frac{d\mu_d^*}{dt}\right). \qquad (2.17)$$

It can be seen that the structure of (2.17) is similar to that of (2.10) and (2.12). For a given amplitude of the excitation field the function $\mu_d^*[H_1(t)]$ may be considered as a time dependent function $\mu_d^*(t)$.

The difference between (2.17) and the equations (2.10) and (2.12) is only in use of the value of differential permeability of the body μ_d^* in (2.17) instead of the conventional permeability of the body μ^*. The value μ_d^* is used in (2.17) in accordance with the definition, as a result of the Taylor expansion of $B(H)$.

In this respect, we wish to emphasize that the expansion of $B(H_\Sigma)$ into a Taylor series for small influences, originally applied in [3], is a necessary operation for a transition from non-linear conversion equations to linear equations with variable coefficients.

This operation is correct not only from the mathematical point of view, but also from the physical.[4] This is confirmed by a good correspondence between calculated results and experimental data, and by obtaining the same expressions in the non-linear equation and parametric approaches. To illustrate the last point we again will use the approximation curve $B(H)$ by the polynomial of the form (1.8). For $H_1(t) = H_m\sin(\omega t)$ we have

$$\mu_d^*[H_1(t)] = \frac{1}{\mu_0}\frac{dB}{dH_1} = \frac{1}{\mu_0}\left(a - \frac{3}{2}bH_m^2 + \underline{\frac{3}{2}bH_m^2\cos(2\omega t)}\right). \qquad (2.18)$$

The underlined term points out the fact, that the second harmonic is contained in the spectrum of the periodic function $\mu_d^*[H_1(t)]$. Due to this, we found from (2.17) for $H_0 = const \neq 0$ the following relation

---

[4]The doubts, that occasionally appear, are related to the fact that the field $H_1(t) = H_m\sin(\omega t)$ periodically takes zero values. However, in the moments when $H_1 = 0$, the function $B(H)$ is expandable into McLaurin series, which is a special case of Taylor series.

2.2. THE INFLUENCE UPON PERMEABILITY

$$u(t) = 6\omega b s w_2 H_0 H_m^2 \sin(2\omega t) \ .$$

The derived expression agrees with (1.18), the latter being evidence of an equivalence between the nonlinear and the parametric approaches, this is true, however, only under certain conditions.

It is obvious that (2.17) compared to (1.18) has a more general character, as it allows to analyze the processes and phenomena in the fluxgates independently of the approximations to the curve $B(H)$.

2.2.3 Auto-Parametric Approach

Some doubts about the validity of expressing the output EMF of a fluxgate via the derivative $dm u_d^*/dt$ are presented in [28, p. 117]. The author of [28] believes that the only correct way to find the EMF is the following:

$$u(t) \sim \frac{dB}{dt} = \frac{dB}{dH}\frac{dH}{dt}\bigg|_{H=H_\Sigma} = \mu_0 \mu_d^*[H_\Sigma(t)]\frac{dH_\Sigma}{dt} \ . \tag{2.19}$$

Here $\mu_d^*[H_\Sigma(t)]$ is the integral field strength: $H_\Sigma = H_1 + H_0$. It is not equal to the function $\mu_d^*[H_1(t)]$ in (2.17).

Let us find the spectrum of the function $\mu_d^*[H_\Sigma(t)]$ when $H_1(t) = H_m \sin(\omega t)$ and the curve $B(H)$ is approximated with the polynomial (1.8):

$$\begin{aligned}\mu_d^*[H_\Sigma(t)] &= \frac{1}{\mu_0}\Big(a - 3bH_0^2 - \underline{6bH_0 H_m \sin(\omega t)} \\ &\quad - \frac{3}{2}bH_m^2 + \frac{3}{2}bH_m^2 \cos(2\omega t)\Big) \ .\end{aligned} \tag{2.20}$$

It can be seen that (2.20) differs from (2.18) by the auxiliary terms related to the field strength H_0.

Taking into account (2.19) and (2.20) we find for the double rod fluxgate (Fig. 1.1b) under $H_0 = const \neq 0$:

$$u(t) = -2s\ w_2 \frac{dB}{dt} = -2s w_2 \mu_0 \mu_d^*[H_\Sigma(t)]\frac{dH_1}{dt} = 6\omega b s w_2 H_0 H_m^2 \sin(2\omega t) \ .$$

The second harmonic of the output EMF appears here as a result of multiplication of the underlined term in (2.20) by

$$dH_1/dt = \omega H_m \cos(\omega t)$$

Again it can be seen that the obtained expression coincides precisely with (1.19) for any values H_m and H_0.

Thus, the work [28], in fact, does not contain a refutation of the correctness of the parametric approach (under the condition $H_\mathrm{m} \gg H_0$), but another approach, namely autoparametric. The analysis of the processes in fluxgates developed in [28] and the proof of its perfect equivalence to the nonlinear approach is given, the latter being one of the advantages of the above mentioned work.

The parametric and autoparametric approaches constitute the basics of modern phenomenological theory of fluxgates, the autoparametric approach being the most useful when considering mechanisms for generation of false signals in fluxgates (see Sect. 3.2).

2.3 Derivation of the Most General Expression for the Output EMF

For quite a time the similar structure of (2.12) and (2.17) suggested the idea of a general theory of fluxgates, which would contain the theories of fluxgates with longitudinal, transversal and other types of core excitations as special cases.

A certain contribution to such a theory was made by Yu. F. Ponomarev [3]. He introduced the concept of a dynamic permeability

$$\mu_\mathrm{dyn}^*$$

(we use a different notation for it in the present work), which he defined for the longitudinal and transversal core excitation as a partial derivative

$$\partial B/(\mu_0 \partial H_0)$$

. If the core remagnetization occurred along the maximum hysteresis loop, or if approximations without hysteresis were used for the cores magnetizing curves (see Sect. 1.3), then it turned out that for the longitudinal excitation

$$\mu_\mathrm{dyn}^* = \mu_\mathrm{d}^*$$

, and for the transversal excitation

$$\mu_\mathrm{dyn}^* = \mu^*$$

. Further analysis may be done within the frameworks of known partial theories [15, 3].

The next step in the development of a general parametric theory was made in [22]. E. N. Langvagen considered the Taylor expansion of the vector function

$$\mathbf{B}(\mathbf{H}_\Sigma)$$

, where

$$\mathbf{H}_\Sigma = \mathbf{H}_1 + H_0$$

2.3. DERIVATION OF THE MOST GENERAL EXPRESSION FOR THE OUTPUT EMF

. The result (for a single core) was the following:

$$\mathbf{B}(\mathbf{H}_\Sigma) = \mu_0 \left[\hat{\mu}^*[\mathbf{H}_1(t)] \cdot \mathbf{H}_1(t) + \hat{\mu}^*_{\text{dyn}}[\mathbf{H}_1(t)] \cdot \mathbf{H}_0(t) \right] . \qquad (2.21)$$

Here

$$\hat{\mu}^*_{\text{dyn}}$$

is the dynamic permeability of the body, with the partial derivatives

$$\partial B/(\mu_0 \partial H_0).$$

If the vector

$$\mathbf{H}_1$$

is aligned with one of the axes of symmetry of the core, for example, with axis

$$\mathbf{x}^0,$$

then the matrix of

$$\hat{\mu}^*_{\text{dyn}}$$

becomes diagonal:

$$\hat{\mu}^*_{\text{dyn}} = \begin{pmatrix} \mu^*_{\text{dyn},x} & 0 & 0 \\ 0 & \mu^*_{\text{dyn},y} & 0 \\ 0 & 0 & \mu^*_{\text{dyn},z} \end{pmatrix}, \qquad (2.22)$$

where

$$\mu^*_{\text{dyn},x} = \mu^*_{d,x},$$
$$\mu^*_{\text{dyn},y} = \mu^*_{d,y}$$

and

$$\mu^*_{\text{dyn},z} = \mu^*_{d,z}$$

are parameters that are already well known to us,
namely the differential and normal body permeabilities.

With this approach (2.11) and (2.16) turn out to be special cases of (2.21), (2.12) and (2.17) being special cases of a more general equation, identical in its structure to (2.10):

$$u(t) = -s_\Sigma w_2 \mu_0 \mathbf{i}^0_w \left(\hat{\mu}^*_{\text{dyn}}[\mathbf{H}_1(t)] \frac{d\mathbf{H}_0}{dt} + \mathbf{H}_0(t) \frac{d\hat{\mu}^*_{\text{dyn}}}{dt} \right), \qquad (2.23)$$

where $u(t)$ is a useful EMF of the fluxgate, s_Σ is the integral cross-sectional area of its cores.

2.3.1 Tensor of the Dynamic Permeability of a Body

Unlike $\hat{\mu}^*$, the tensor $\hat{\mu}^*_{\text{dyn}}$ describes not only the anisotropy of the material and the shape of the body (see Sect. 1.4), but also the anisotropy induced by the drive field $\mathbf{H}_1(t)$. This anisotropy can be observed only in presence of a measured field $\mathbf{H}_0$.

By analogy with the expression (1.37) $\hat{\mu}^*_{\text{dyn}}$ may be defined as follows

$$\hat{\mu}^*_{\text{dyn}} = \frac{1}{\mu_0} \frac{dB}{d\mathbf{H}_1} = \hat{\mu}^* + \mathbf{H}_1 \frac{d\hat{\mu}^*}{d\mathbf{H}_1} \ . \tag{2.24}$$

For $\mathbf{H}_1 \| \mathbf{x}^0$ in accordance with (2.22) we find that with the longitudinal excitation in (2.24) both terms are required, and as a result $\mu^*_{\text{dyn},x} = \mu^*_{\text{d},x}$. In the case of transversal excitation only the first term is required, as the projection of vector $\mathbf{H}_1$ onto axes $\mathbf{y}^0$ and $\mathbf{z}^0$ are equal to zero, hence $\mu^*_{\text{dyn},y} = \mu^*_{\text{d},y}$ and $\mu^*_{\text{dyn},z} = \mu^*_{\text{d},z}$.

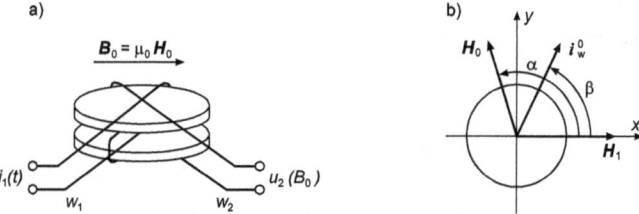

Figure 2.1: Fluxgate with disk cores (a) and the vector diagram explaining its operation (b).

It is convenient to demonstrate the properties of the following model of the fluxgate which has a disk core (Fig. 2.1). Let us assume that the material of the disks is isotropic and that the vectors $\mathbf{H}_1$ and $\mathbf{H}_0$, as well as the axes of the drive and sense coils, lie in the disk plane. In this case the tensor $\hat{\mu}^*$ degenerates into a scalar, while tensor $\hat{\mu}^*_{\text{dyn}}$ characterizes only the anisotropy induced by the excitation field $\mathbf{H}_1(t)$.

For thin disks, one may neglect the component $\mu^*_{\text{dyn},z}$ orientated along the normal, with respect to the disks, as, in this case, $\mu^*_{\text{dyn},z} \to 1$, and $\mu^*_{\text{dyn},x} \gg 1$, $\mu^*_{\text{dyn},y} \gg 1$. Hence, instead of (2.22) we have

$$\hat{\mu}^*_{\text{dyn}} = \begin{vmatrix} \mu^*_{\text{dyn},x} & 0 \\ 0 & \mu^*_{\text{dyn},y} \end{vmatrix}, \tag{2.25}$$

where $\mu^*_{\text{dyn},x} = \mu^*_{\text{d}}$ and $\mu^*_{\text{dyn},y} = \mu^*$. The difference $H_1 d\mu^*/dH_1 = \mu^*_{\text{d}} - \mu^*$ between these parameters, derived from (1.37), is exactly the induced anisotropy coefficient.

Defining the angles $\alpha \mathbf{H}_1, \mathbf{H}_0$ and $\beta \mathbf{H}_1, \mathbf{i}^0_w$, where $\mathbf{i}^0_w$ is the axis of the sense coil (see Fig. 2.1b), we obtain from (2.23), for $H_0 = const \neq 0$:

2.3. DERIVATION OF THE MOST GENERAL EXPRESSION FOR THE OUTPUT EMF

$$u(t) = -s_\Sigma w_2 \mu_0 H_0 \frac{d}{dt}(\mu_d^* \cos\alpha \cos\beta + \mu^* \sin\alpha \sin\beta)$$

$$= -s_\Sigma w_2 \mu_0 H_0 \frac{d}{dt} \begin{vmatrix} \mu_d^* & -\sin\alpha \sin\beta \\ \mu^* & \cos\alpha \cos\beta \end{vmatrix}. \quad (2.26)$$

(2.26)

The expression (2.26) permits to find the useful EMF for a fluxgate with an arbitrary mutual orientation of vectors $\mathbf{H}_1$, $\mathbf{H}_0$ and $\mathbf{i}_w^0$.

2.3.2 Special Cases of Excitation

Longitudinal Drive Field ($\mathbf{H}_1 \parallel \mathbf{H}_0$)

For this case from (??) one has

$$u(t) = -s_\Sigma w_2 \mu_0 H_0 \cos\beta \frac{d\mu_d^*}{dt}. \quad (2.27)$$

As $\mu_d^*(H)$ is even and $H_1(t)$, as a rule, is odd, the periodic function $\mu_d^*[H_1(t)]$ is expandable into a Fourier series with respect to even harmonics [3]:

$$\mu_d^*[H_1(t)] = \mu_{d,0}^* + \sum_{n=1}^{\infty} \mu_{d,2n}^* \cos(2n\omega t), \quad (2.28)$$

where

$$\mu_{d,0}^* = \frac{1}{\pi} \int_0^\pi \mu_d^*(\omega t)\, d(\omega t), \quad (2.29)$$

$$\mu_{d,2n}^* = \frac{2}{\pi} \int_0^\pi \mu_d^*(\omega t) \cos(2n\omega t)\, d(\omega t) \quad (2.30)$$

are the constant components and amplitudes of the even harmonics respectively, and $n = 1, 2, 3, \ldots$ are integers.

Substituting (2.28) into (2.27) one may conclude that the spectrum of the useful EMF will also contain only even harmonics. This also follows from the graphical representation (see Fig. 2.2).

Transversal Drive Field ($H_1 \perp H_0$)

For this case one has, from (2.26), the following:

$$u(t) = -s_\Sigma w_2 \mu_0 H_0 \sin\beta \, \frac{d\mu^*}{dt} \,. \tag{2.31}$$

Here also the function $\mu^*(H)$ is even, and the function $H_1(t)$ is odd.

Hence, in an analogy with (2.28)–(2.30), for the periodic function $\mu^*[H_1(t)]$ one may write the following expression:

$$\mu^*[H_1(t)] = \mu_0^* + \sum_{n=1}^{\infty} \mu_{2n}^* \cos(2n\omega t) \,, \tag{2.32}$$

where

$$\mu_0^* = \frac{1}{\pi} \int_0^\pi \mu^*(\omega t) \, d(\omega t) \,, \tag{2.33}$$

$$\mu_{2n}^* = \frac{2}{\pi} \int_0^\pi \mu^*(\omega t) \, \cos(2n\omega t) \, d(\omega t) \,. \tag{2.34}$$

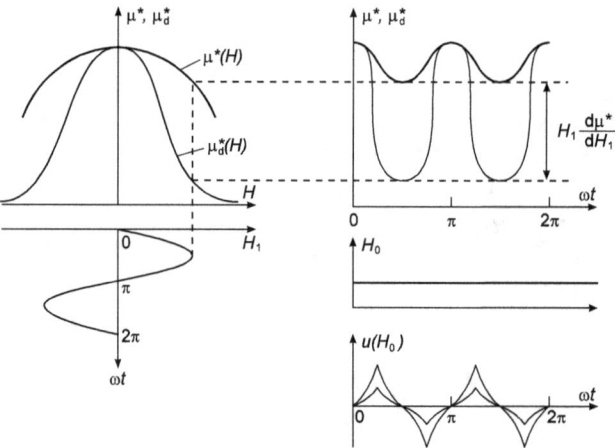

Figure 2.2: Analysis of the fluxgate performance for various excitation methods.

It is obvious that in this case the spectrum of the output EMF of fluxgate will also contain only even harmonics (see Fig. 2.2). The amplitudes μ_{2n}^* are always less than amplitudes $\mu_{d,2n}^*$. Therefore only for a rectangular excitation wave and a deep saturation of the cores the difference between the amplitudes will be minimal.

2.3. DERIVATION OF THE MOST GENERAL EXPRESSION FOR THE OUTPUT EMF

Rotating Drive Field($\mathbf{H}_1 = f(t)$, $|\mathbf{H}_1| = const \neq 0$)

In case of fluxgates, this way of excitation was proposed by E. N. Langvagen and it follows from his concept that an anisotropy induced in the core is induced by the field $\mathbf{H}_1$.

The expression for the output EMF of the fluxgate may be obtained from (2.26) by replacing the fixed reference system $x0y$ (Fig. 2.1b) with a rotating one

$$\begin{aligned}u(t) &= -s_\Sigma w_2 \mu_0 H_0 \frac{d}{dt}\left(\frac{\mu_d^* + \mu^*}{2}\cos\varphi + \frac{\mu_d^* - \mu^*}{2}\cos(2\omega t - \varphi)\right) \\ &= \omega s_\Sigma w_2 \mu_0 H_0 H_1 \frac{d\mu^*}{dH_1}\sin(2\omega t - \varphi)\end{aligned} \quad (2.35)$$

where $\varphi = (\alpha - \beta)$ is the initial phase of the EMF. One may conclude from these expressions:

1. With a rotating drive field the difference $\mu_d^* - \mu^* = H_1 d\mu^*/dH_1$, i.e. the coefficient of the induced anisotropy, may serve as the variable parameter (see Fig. 2.2);

2. The output EMF contains only the second harmonic[5]

3. The second harmonic amplitude of EMF is proportional to the modulus $|\mathbf{H}_0|$, the information on the direction of $\mathbf{H}_0$ is provided by the initial phase φ.

2.3.3 Calculation of the Transformation Ratio

By comparison of the obtained expressions, one may conclude that in the frameworks of the parametric approach, the useful EMF for any fluxgate may be found, in fact, in two steps.

First one finds the dependencies $\mu_d^*(H_1)$, $\mu^*(H_1)$ and $H_1 d\mu^*/dH_1$, corresponding to the way of core excitation. These dependencies are presented in table 2.1.

Second, taking the shape of waves or peculiarities of the field polarization $\mathbf{H}_1(t)$ into account, the mentioned dependencies are expanded into Fourier series and the series coefficients, i.e. the amplitudes of the even harmonics $\mu_{d,2n}^*$, μ_{2n}^* and $(H_1 d\mu^*/dH_1)_{2n}$ are found. These values eventually turn out to be proportional to the amplitudes of the even harmonics of the useful EMF, and, consequently, the

---

[5]This holds true for circular polarization of the field $\mathbf{H}(t)$. In case of an elliptical polarization in the EMF the higher even harmonics will appear. In the limit, ie. when one of the components of the rotating field tends to zero, we come, of course, to the longitudinal or transversal (depending on the orientation of the vector $\mathbf{H}_0$) method of core excitation.

{Magnetization curve B(H)	Parameter		
	μ_d^*	μ^*	$H\dfrac{d\mu^*}{dH} = \mu_d^* - \mu^*$
$B = aH - bH^3$	$\dfrac{1}{\mu_0}(a - 3bH^2)$	$\dfrac{1}{\mu_0}(a - bH^2)$	$-\dfrac{1}{\mu_0} 2bH^2$
$B = \alpha \arctan(\beta H)$, $\alpha = \dfrac{2}{\pi} B_S$, $\beta = \dfrac{\pi}{2}\dfrac{1}{H_S}$	$\dfrac{1}{\mu_0}\dfrac{\alpha\beta}{1+\beta^2 H^2}$	$\dfrac{1}{\mu_0}\dfrac{\alpha \arctan(\beta H)}{H}$	$\dfrac{1}{\mu_0}\alpha\dfrac{\dfrac{\beta H}{1+\beta^2 H^2} - \arctan(\beta H)}{H}$
$B = aH - b\sinh H$	$\dfrac{1}{\mu_0}(a - b\cosh H)$	$\dfrac{1}{\mu_0}\left(a - \dfrac{b\sinh H}{H}\right)$	$\dfrac{1}{\mu_0}\dfrac{bH\cosh H + b\sinh H}{H}$
$B = a\tanh H$	$\dfrac{1}{\mu_0}\dfrac{a}{\cosh^2 H}$	$\dfrac{1}{\mu_0}\dfrac{a\tanh H}{H}$	$\dfrac{1}{\mu_0}\dfrac{\dfrac{aH}{\cosh^2 H} - a\tanh H}{H}$
$B = aHe^{-b^2 H^2}$	$\{\dfrac{1}{\mu_0}ae^{-b^2 H^2} \times (1 - 2b^2 H^2)$	$\dfrac{1}{\mu_0}ae^{-b^2 H^2}$	$-\dfrac{1}{\mu_0} 2b^2 H^2 a e^{-b^2 H^2}$

Table 2.1: Approximating functions of the magnetization curve and core parameters

same is true for the corresponding transformation ratio of the fluxgate defined via the derivative

$$G_{2n} = \frac{dE_{m,2n}}{dB_{0,i}}, \qquad (2.36)$$

where $E_{m,2n}$ is the amplitude of the corresponding even harmonics, and $B_{0,i} = \mu_0 H_{0,i}$ is the measured component of $\mathbf{B}_0$ (see Sect. 2.1).

Under the same conditions for fluxgates with the longitudinal, rotating, and transversal drive field, we have the following relation for the transformation ratios for the second harmonic ($n = 1$):

$$G_{2,\parallel} : G_2 : G_{2,\perp} = \mu_{d,2}^* : \left(H_1\frac{d\mu^*}{dH_1}\right) : \mu_2^*. \qquad (2.37)$$

For the simplest approximation of the form (1.8), used for the magnetization

2.3. DERIVATION OF THE MOST GENERAL EXPRESSION FOR THE OUTPUT EMF

curve of the cores, the ratio of numbers corresponding to (2.37) is 3 : 2 : 1, that may be seen from the first row of table 2.1.

2.3.4 Calculation of the Misbalance EMF

The expression (2.26) is not complete in the sense that only the useful EMF may be found from it.

However, as was shown before (see Sect. 1.2.4), at the fluxgate output there is always a so-called misbalance EMF, which does not convey information about the field to be measured, yet which, being an additional signal in the measurement amplitude, interferes with the process of recovery of a weak useful EMF. The misbalance EMF $u^{\neq}t$ may be represented as the difference of the two transformation EMFs $u^{\neq}(t) = u'(t) - u''(t)$, induced in the sense coil of the fluxgate by the two opposing direction magnetic fluxes, created by the drive field. The greater the mismatch between the cores or their halves, and their windings, the bigger the misbalance EMF becomes. Usually, $|u^{\neq}|$ is less than $|u'|$ and $|u''|$ by hundreds of times, hence one may assume that $|u'| \approx |u''|$. Introducing misbalance coefficient [6]

$$\varepsilon = \overline{\varepsilon}.$$

$$\varepsilon = \frac{|u^{\neq}|}{|u'|+|u''|} \approx \frac{|u^{\neq}|}{2|u'|}$$

and differentiating the first term of (2.21) with respect to time, then taking into consideration the notations used above, we find

$$u^{\neq}(t) = -s_\Sigma w_2 \mu_0 \varepsilon \cdot \left(\mu^* \cos\beta \frac{dH_1}{dt} + H_1 \cos\beta \frac{dm u^*}{dt} + \mu^* H_1 \frac{d}{dt}\cos\beta\right)$$

where

$$s = s_\Sigma/2$$

is the area of the cross-section of one core or one of its halves.

Representing the derivative $d\mu^*/dt = (d\mu^*/dH_1)(dH_1/dt)$ and taking into account (1.37), we finally obtain:

$$\begin{aligned}
\dot{u}^{\neq}(t) &= -s_\Sigma w_2 \mu_0 \varepsilon \left(\mu_d^* \frac{dH_1}{dt}\cos\beta + \mu^* H_1 \frac{d}{dt}\cos\beta\right) \\
&= -s_\Sigma w_2 \mu_0 \varepsilon \begin{vmatrix} \mu_d^* & -H_1 \frac{d}{dt}\cos\beta \\ \mu^* & \frac{dH_1}{dt}\cos\beta \end{vmatrix}.
\end{aligned} \quad (2.38)$$

---

[6] In Sect. 3.2 it will be shown that the misbalance coefficient should be introduced separately for each harmonic. Here we mean some average value of the misbalance coefficient

Longitudinal or Transversal Excitation Field ($\cos\beta = const$)

For this case we have

$$u^{\neq}(t) = -s_{\Sigma}w_2\mu_0\varepsilon\mu_d^*[H_1(t)]\frac{dH_1}{dt}\cos\beta .\qquad(2.39)$$

As the periodic function $\mu_d^*[H_1(t)]$, in accordance with (2.28), contains only even harmonics, and the function $H_1(t)$ contains only odd ones, the misbalance EMF spectrum will contain only even harmonics. Thus, for the sinusoidal drive field, $H_1(t) = H_m \sin(\omega t)$, we obtain

$$u^{\neq}(t) = -\omega s_{\Sigma}w_2\varepsilon\mu_0 H_m \cos\beta \left(\frac{2\mu_{d,0}^* + \mu_{d,2}^*}{2}\cos(\omega t) + \sum_{n=1}^{\infty}\frac{\mu_{d,2n}^* + \mu_{d,(2n+2)}^*}{2}\cos[(2n+1)\omega t]\right)$$

Attention should be drawn to the fact that the odd EMF misbalance harmonics are expressed here via the amplitudes of the even harmonics of the cores' differential permeability.

Rotating excitation field ($H_1 = const \neq 0, \cos\beta = \cos(\omega t - \varphi)$)

From expression (2.38) we have

$$u^{\neq}(t) = \omega s_{\Sigma}w_2\mu_0\varepsilon\mu^* H_1 \sin(\omega t - \varphi) .\qquad(2.40)$$

It can be seen that the misbalance EMF in this case contains only the main frequency (the fundamental) and that the value of this EMF, under the same conditions, is proportional to $\mu^* = \mu^*(H_1)$, ie. to the normal permeability of the cores.

2.3.5 A General Expression for the Output EMF

Now, on the basis of (2.26) and (2.38), the complete expression for the output EMF of the fluxgate may be written as:

$$u_{\Sigma}(t) = -s_{\Sigma}w_2\xi\mu_0\left(\varepsilon\begin{vmatrix}\mu_d^* & -H_1\frac{d}{dt}\cos\beta \\ \mu^* & \frac{dH_1}{dt}\cos\beta\end{vmatrix} + \frac{d}{dt}H_0\begin{vmatrix}\mu_d^* & -\sin\alpha\sin\beta \\ \mu^* & \cos\alpha\cos\beta\end{vmatrix}\right).$$
$$(2.41)$$

Here, the first of the determinants gives the spectrum of the misbalance EMF, consisting of odd harmonics, the second determinant gives the useful signal spectrum, consisting of even harmonics. In the above expression the value of the measured

2.3. DERIVATION OF THE MOST GENERAL EXPRESSION FOR THE OUTPUT EMF

field H_0 is put into the body of the differential equation, as in the general case $H_0 \neq const$. The common multiplier contains the coefficient $\xi \leq 1$, which depends upon the mutual position of the cores and the windings of the fluxgate (see Sect. 2.4).

Expression (2.41) is obviously the most general one. It is suitable for fluxgate operation analysis under any way of core excitation, and the cores themselves may either be isotropic (see Fig. 2.1), or anisotropic.

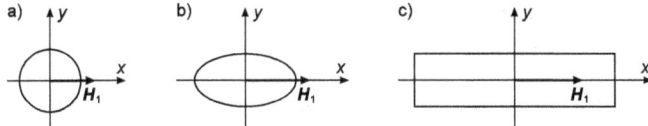

Figure 2.3: A scheme of a change from isotropic cores (a) to anisotropic ones (b and c).

A scheme of a change from isotropic to anisotropic cores is depicted in Fig. 2.3 where a disk, an ellipsoid and a rectangular plate are shown. If the cores are thin enough (this condition is almost always fulfilled, as the cores are made from a permalloy tape of industrial production), and if the axes of symmetry of cores are taken as the coordinate axis, then for all of them a square matrix of the form (2.25) of the dynamic permeability tensor $\hat{\mu}^*_{\text{dyn}}$ remains valid, and for the corresponding construction of the fluxgate the most general expression (2.41) remains valid, too. One thing should be remembered: the core shape anisotropy leads to a loss of alignment between the external magnetizing force and the internal one (see Sect. 1.4.1). This is why in the case of anisotropic cores, the conditions for imposing longitudinal, transversal and rotating excitation fields, respectively, will not be the conditions listed in Sect. 2.3.2 (which are valid only for isotropic cores), but the most general ones:

$$\mathbf{h}_1 \| \mathbf{h}_0, \quad \mathbf{h}_1 \perp \mathbf{h}_0, \quad \mathbf{h}_1(t), |\mathbf{h}_1| = const \neq 0 . \qquad (2.42)$$

Here $\mathbf{h}_0$ and $\mathbf{h}_1$ are the vectors of magnetizing force of the inner drive field and that of the magnetic field to be measured.

Note, that in case of anisotropic cores the condition $\mathbf{h}_0 \| \mathbf{h}_1$ may be practically fulfilled also if the external drive field and the measured field are mutually orthogonal. This is true, for example, for a long thin rod, oriented in such a way that its longitudinal axis is not perpendicular to $\mathbf{H}_1$ or $\mathbf{H}_0$. For the rotating drive field the anisotropy of the core shape leads to the following effect: under circular polarization of the inner field, that corresponds to $|\mathbf{h}_1| = const \neq 0$, the external field has to be elliptic in case of the core of elliptical shape (see Fig. 2.3b) and vice versa.

The expressions (2.41) and (2.42) are the most general ones in the sense that they are applicable to the analysis of open core fluxgates, as well as to closed core

CHAPTER 2. FUNDAMENTALS OF PARAMETRIC THEORY

fluxgates. In the latter case, in accordance with the discussion in Sect. 1.4, in the first determinant of (2.41) the body parameters μ_d^* and μ^* are to be replaced by the material parameters, μ_d and μ, because, with respect to the circular drive field $h_1(t) \equiv H_1(t)$, the core demagnetization coefficient is equal to zero.

2.4 Open Core Fluxgates

The most popular and experimentally well proved fluxgates of this type are fluxgates which implement the method of the longitudinal excitation (see Figs. 1.1b, 1.3 and 1.5).

Accumulated experience and special experiments [3, 35, 47, 71] have shown that the piecewise linear approximation of the mean magnetizing curve $B(H)$ is quite adequate for rod cores which are widely used.

The graphs given in Fig. 2.4 are typical for the piecewise linear approximation to the mean magnetizing curve and other parameters of rod-type cores. The solid lines show dependencies and periodic processes related to the mean magnetizing curve, the dashed lines shows the dependencies and processes, which are related to the maximum hysteresis loop of the cores (see Sect. 1.3 and Fig. 1.6). It can be seen

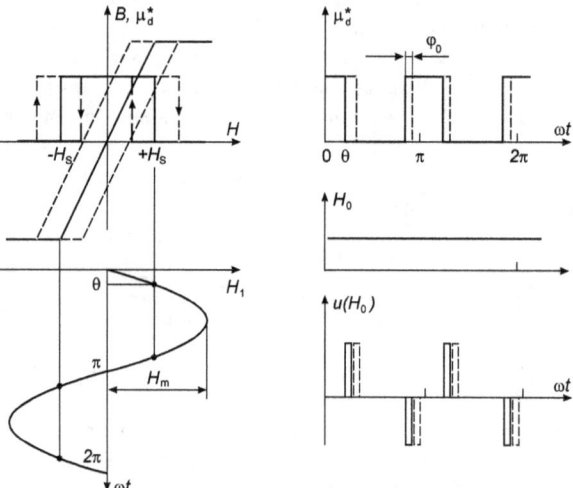

Figure 2.4: Analysis of the fluxgate performance under piecewise linear approximation of the magnetizing curve of the rod cores in the given magnetizing force of the drive field.

from the figure, that the periodic function $\mu_d^*[H_1(t)]$ is completely characterized by the values $\mu_{d,\max}^*$, $\mu_{d,\min}^*$ and $\omega t = 0$. Also, it may be seen that when a transition

2.4. OPEN CORE FLUXGATES

from mean magnetizing curve to the limiting hysteresis loop of the cores occurs, the pattern of the periodic process does not change, being only shifted to the right along axis ωt by the angle φ_0. The same property is also observed for the periodic function $u(t)$, i.e. the useful EMF of the fluxgate.

In accordance with the discussion in Sect. 1.4, for the chosen approximation we have

$$\mu^*_{d,\,max} = \frac{\mu_{d,\,max}}{1+N(\mu_{d,\,max}-1)} = \frac{\mu_{d,\,max}\,m}{m+\mu_{d,\,max}-1}\bigg|_{\mu_{d,\,max} \gg m \gg 1} \approx m \;,$$

$$\mu^*_{d,\,min} = \frac{\mu_{d,\,min}}{1+N(\mu_{d,\,min}-1)}\bigg|_{\mu_{d,\,min} \to 1} \approx 1$$

Taking into consideration the obtained value we find from (2.29) and (2.30)[7]

$$\mu^*_{d,0} = \frac{2m\theta}{\pi} \;, \tag{2.43}$$

$$\mu^*_{d,2n} = \frac{2m}{n\pi}\sin(2n\theta) \;,$$

where θ is the so-called angle of core saturation (see Sect. 2.6.1).

The series, in accordance with (2.28), will have the following form [3]

$$\mu^*_d[H_1(t)] = \frac{2m}{\pi}\left(\theta + \sum_{n=1}^{\infty} \frac{\sin(2n\theta)}{n}\cos(2n\omega t)\right) \;. \tag{2.44}$$

It may be demonstrated that when the numeric values of the coefficients of the series (2.44) are applied to the approximation of the limiting hysteresis loop the numerical values of the coefficients $\mu^*_{d,0}$ and $\mu^*_{d,2n}$ will remain unchanged, and only the initial phase φ_{2n} will be added, which is dependent on the coercive force of the dynamic loop.

Substituting (2.44) into (2.41) one may find the complete spectrum of the output EMF of the fluxgate. For the EMF of the useful signal assuming $H_0 = const \neq 0$ and $\beta \approx 0$ (the axis of the measuring coil coincides with the longitudinal axes of the cores, see Figs. 1.1b, 1.3, 1.5) we find

$$\begin{aligned} u_{2n}(t) &= -s_\Sigma w_2 \xi \mu_0 H_0 \cos\alpha \frac{d}{dt}\left(\mu^*_d[H_1(t)]\right) = \\ &= \frac{4}{\pi}\omega s_\Sigma w_2 \xi m \sin(2n\theta)\mu_0 H_0 \cos\alpha \sin(2n\omega t) \;. \end{aligned} \tag{2.45}$$

---

[7]Under the approximation of the mean magnetizing curve the function $\mu^*_d(\omega t)$ is evenly symmetrical in the range $0 - \pi$. Due to this the integration is done from 0 to $\pi/2$, i.e. from 0 to θ, and then the obtained result is doubled.

Taking into account, that $\mu_0 H_0 \cos\alpha = B_{0,i}$, we derive in accordance with (2.36), the following expression for the fluxgate transformation ratio

$$G_{2n} = \frac{4}{\pi}\omega s_\Sigma w_2 \xi m \sin(2n\theta) \ . \tag{2.46}$$

Hence, it follows that the transformation ratio for every even harmonic of the output EMF depends on the angle θ, all other conditions being fixed,

As it will be shown below, it is always possible to choose such a mode that the modulus $|\sin(2n\theta)|_{\max} = 1$, and consequently, the maximal value of the fluxgate transformation ratio for any even harmonics of EMF will be equal to:

$$G_{\max} = \frac{4}{\pi}\omega s_\Sigma w_2 \xi m \ . \tag{2.47}$$

The saturation angle θ of the cores is a function of the core parameter and the drive field.

For the chosen piecewise linear approximation of the mean magnetizing curve, the angle θ is uniquely defined by the values H_S and $H_{1,\max}$, where H_S is the magnetizing force of the core saturation field, and $H_{1,\max}$ is the maximum force of the drive field (for sinusoidal field one has $H_{1,\max} = H_m$, see Fig. 2.4).

For the three most frequently applied fields $H_1(t)$, ie. for sinusoidal, triangular and rectangular fields, we have

$$\theta_{\sin} = \arcsin\frac{H_S}{H_m}, \quad \theta_{\text{tr}} = \frac{\pi}{2}\frac{H_S}{H_{\max}}, \quad \theta_{\text{rec}} = \frac{\pi - \tau}{2}, \tag{2.48}$$

where τ is the period of the drive field.

Substituting (2.48) into (2.46) we find the corresponding expressions for the fluxgates' transformation ratios for the second harmonic:

$$G_{2,\sin} = G_{\max}\frac{2H_S}{H_m}\sqrt{1 - \left(\frac{H_S}{H_m}\right)^2},$$

$$G_{2,\text{tr}} = G_{\max}\sin\left(\pi\frac{H_S}{H_{\max}}\right), \tag{2.49}$$

$$G_{2,\text{rec}}|_{H_{\max} > H_S} = G_{\max}\sin(\pi - \tau) \ ,$$

where $G_{\max}$ is a constant coefficient defined by (2.47).

Using the expressions (2.49) it is also possible to obtain the corresponding dependencies $G_2(H_{\max})$. These dependencies, normalized with respect to the relative values $G_2/G_{\max}$ and $H_{\max}/H_S$, are shown in the Fig. 2.5. It can be seen that the conversion coefficients $G_{2,\sin}$, $G_{2,\text{tr}}$ and $G_{2,\text{rec}}$ achieve their maximum value for different values of $H_{\max}/H_S$. The corresponding relations, which provide the maxima for the transformation ratios for the second harmonic, are as follows:

$$(H_m)_{\text{opt}} = \sqrt{2}\,H_S \ ,$$

2.4. OPEN CORE FLUXGATES

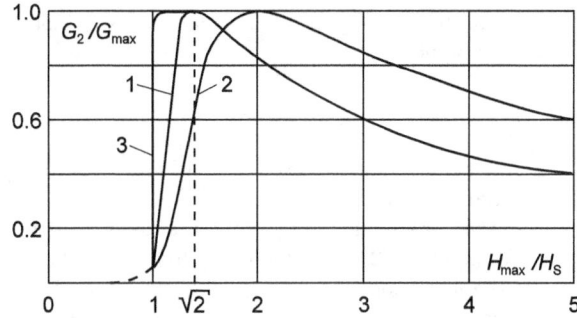

Figure 2.5: The dependence of the fluxgate transformation ratios upon the degree of the relative (over) excitation of cores. 1 – sinusoidal excitation signal; 2 – triangular excitation signal; 3 – rectangular excitation signal.

$$(H_{\max})_{\mathrm{opt}} = 2H_{\mathrm{S}}, \qquad (2.50)$$

$$(H_{\max})_{\mathrm{opt}} > H_{\mathrm{S}}, \quad \tau_{\mathrm{opt}} = \pi/2 .$$

For the pulse-rectangular drive field the angle θ is defined only by the pulse duration $\theta = (\pi - \tau)/2$. This is why, starting from $(H_{\max})_{\mathrm{opt}} > H_{\mathrm{S}}$, the transformation ratio $G_{2,\mathrm{rec}}$, as well as the ratios for higher even harmonics $G_{2n,\mathrm{rec}}$, are not dependent further on the growth of $H_{\max}$. This fact is important in practice, as, for any other shape of the drive field, the instability of the drive field's maximum (the amplitude in case of a sinusoidal field) is the main contributor to instability of the fluxgate transformation ratio.

Here we also note, that, in analogy to (2.49), formulae for fluxgate ratios for higher even harmonics also may be obtained, though the use of fluxgates on higher harmonics is not widely spread.

In this sense (2.46) is universal. Within the frameworks of a chosen approximation for the mean magnetizing curve this formula remains valid for any shape of the excitation field.

2.4.1 Voltage Mode Excitation

The fluxgate performance was studied above for the mode of a given drive current. As the given current mode may be achieved either by the use of a generator with large inner resistance, or by placement of a resistor in the excitation circuit, with the ohmic resistance prevailing over the inductive resistance of the fluxgate drive coil (due to this, vast losses of energy are unavoidable), in some cases the regime of a given excitation voltage, ie. the regime of a given induction of the excitation field, is preferred.

The given voltage mode is provided by connection of the drive coil to a low inner resistance generator, under the condition that the ohmic resistance of the winding is negligibly small. If the integral ohmic resistance is so small, that the voltage drop across it may be neglected, then

$$u_1(t) = u_L(t) = w_1 s \frac{dB_1}{dt} \qquad (2.51)$$

where u_L is a voltage gradient across the excitation winding with w_1 being the number of turns sequentially encompassing both cores (see Figs. 1.1b, 1.3, 1.5), the value s is the area of the cross-section of one core, the value B_1 is the magnetic induction in the cores. Integrating (2.51) and taking into account the absence of a constant component of the induction (the magnetizing curve is odd-symmetrical, and there is neither a constant component nor even harmonics in the voltage $u_1(t)$), we have

$$B_1(t) = \frac{1}{w_1 s} \int u_1(t)\, dt \ . \qquad (2.52)$$

Let us show that the parametric approach is perfectly applicable for analysis of the fluxgate performance in a given drive field induction mode.

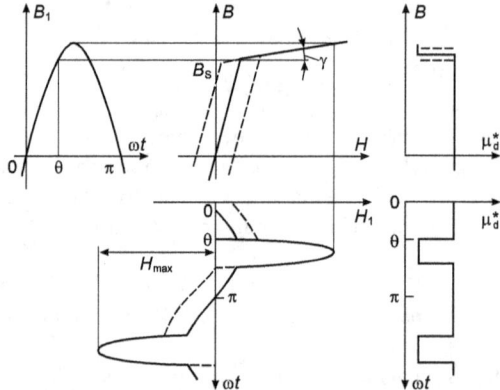

Figure 2.6: Analysis of the fluxgate performance under piecewise linear approximation of the magnetizing curve for rod-type cores in the mode of given induction of the drive field.

The required graphs are shown in the Fig. 2.6. If the given voltage changes are governed by the law $u_1(t) = U_m \cos(\omega t)$, then in accordance with (2.52) we have

$$B_1(t) = B_m \sin(\omega t)$$

where $B_m = U_m/(\omega w_1 s)$, that is the shape of magnetic induction wave in the cores will be a sinusoidal one. However, due to the nonlinearity of the curve $B(H)$ the

2.4. OPEN CORE FLUXGATES

current $i_1(t)$ and, consequently, the force of the drive field $H_1(t)$ will be significantly different from the sinusoidal form.

Yet, one may ignore the shape of the field force $H_1(t)$, which previously has been taken into consideration in order to find the spectral composition of the periodic function $\mu_d^*[H_1(t)]$, and immediately try to find out the spectral composition of the periodic function $\mu_d^*[B_1(t)]$.

Indeed, the parameter μ_d^*, as well as parameter μ^*, may be considered not only as dependent on H_1, but also on B_1. The corresponding graphs are shown in Fig. 2.6. The solid lines denote the dependencies and the periodic processes, related to the mean curve of core magnetization, the hatched lines show the same parameters, related to the limiting hysteresis loop.

It can be seen, that the essentially non-sinusoidal shape of the field force $H_1(t)$ corresponds to the sinusoidal shape of the magnetic induction wave $B_1(t)$. As far as the character of dependencies $\mu_d^*(B_1)$ and $\mu_d^*[B_1(t)] = \mu^*(\omega t)$ is concerned, they are quite analogous to the dependencies $\mu_d^*(H_1)$ and $\mu_d^*[H_1(t)] = \mu_d^*(\omega t)$ shown in Fig. 2.4.

The latter means that for the given drive field induction one may use the saturation angle θ. In our case, when the angle γ is small and $B > B_S$, the coefficients $\mu_{d,0}^*$ and $\mu_{d,2n}^*$ may be calculated from (2.43a) and (2.43b), and the conversion coefficient or the fluxgate may be defined with (2.46). It is necessary, though, to remember that in the given induction mode the angle θ turns out to be a function of the ratios B_S/B_m, or $B_S/B_{\max}$.

In particular, for the sinusoidal and triangular shapes of the induction wave $B_1(t)$ we derive, in analogy to (2.48) and (2.49) the following:

$$\theta_{\sin} = \arcsin \frac{B_S}{B_m}, \quad \theta_{\text{tr}} = \frac{\pi}{2} \frac{B_S}{B_{\max}}, \tag{2.53}$$

$$G_{2,\sin} = G_{\max} \frac{2 B_S}{B_m} \sqrt{1 - \left(\frac{B_S}{B_m}\right)^2}, \quad G_{2,\text{tr}} = G_{\max} \sin\left(\pi \frac{B_S}{B_{\max}}\right). \tag{2.54}$$

Note, that in order to obtain the triangular shape of the induction wave $B_1(t)$, the shape of the excitation voltage $u_1(t)$, applied to the fluxgate in correspondence with (2.52), should be rectangular.

Using (2.54) it is possible to find out the dependencies $G_2(B_{\max})$. They will be identical to the dependencies shown in Fig. 2.5, if the relative values $B_{\max}/B_S$ are plotted along the abscissa.

It can also be seen from Fig. 2.6 that in the process of transition from the mean magnetizing curve to the maximum hysteresis loop, the function $\mu_d^*(B_1)$ changes insignificantly, and the changes become even smaller with the reduction of the angle γ (see the changes of the function $\mu_d^*(H_1)$ in Fig. 2.4). Consequently, the changes of the values of the initial phases φ_{2n} of the output EMF $u_{2n}(t)$ will be small in the same degree, if for some reason, for instance due to a deviation of the frequency of the drive generator, the dynamic coercive force H_f of the fluxgate cores starts to change.

The latter consideration is very important in practice, as a high level of phase stability of the fluxgate output signal achieved in a given induction mode, guarantees a high stability of the zero readings of the fluxgate tools.

Thus, at least for the sinusoidal and triangular wave shapes, it is possible to see the doubtless advantages of a given induction mode with respect to the mode of a given magnetizing force of the drive field of the fluxgates.

2.4.2 Additional Data for Transformation Ratio Calculation

Now we will discuss the problem of calculating the fluxgate transformation ratio G_{max} defined by (2.47). The latter equation is the term, entering the coefficients of the formulas (2.49) and (2.54), and consequently it is independent of the choice of the mode of operation with respect to the excitation circuit.

The coefficients ω, s_Σ and w_2 used in (2.47) are known prior to calculation. The parameter m is estimated using (1.53) and (1.54) if the length of the fluxgate measuring coil is comparable to its core length. Hence, only formulae for calculating the coefficient $\xi \leq 1$ are still to be found.

ξ may be represented as the product of three coefficients [3]:

$$\xi = \xi_1 \xi_2 \xi_3 \,, \tag{2.55}$$

where ξ_1 is a coefficient representing the distance of the two rods in a double rod fluxgate, ξ_2 is a factor depending on the ratio of the average diameter of the sense coil to the length of the rod cores and ξ_3 is a coefficient dependent upon the shunt effect of the core and other metallic parts of the fluxgate, which form short cut turns with respect to the sense coil.

ξ_1 may be calculated using the following formula [3, 35, 36]

$$\xi_1 = \frac{m_2}{m_1} = \frac{2 - e^{-ax}}{2} \,, \tag{2.56}$$

where $a = 1/\{2\delta \ln[(kl)/(\delta + \rho) - 1]\}$; k is a coefficient taken from table 1.1, the parameters l, δ and ρ are the same as in (1.57), x is the distance between the central planes and lines of the cores.

The physical meaning of ξ_1 is obvious. When $x = 0$ (the cores touch each other along their full length), then $e^{-ax} = 1$ and $\xi_1 = 0.5$, i.e. the permeability of the system consisting of two cores, m_2, is half the permeability of one core, m_1. This conclusion is in agreement with (1.57), as the total cross-section area for the cores becomes equal to $s_\Sigma = 2s$. Under $x \to \infty$ the value of $e^{-ax} \to 0$, hence $\xi_1 = 1$. Consequently, ξ_1 is kept within the boundaries $0.5 < \xi_1 < 1$.

ξ_2 is calculated in accordance with the following formula

$$\xi_2 = \frac{1}{\sqrt{1.5 \left(\frac{d_m}{l}\right)^2 + 1}} \,, \tag{2.57}$$

2.5. CLOSED CORE FLUXGATES

where d_m is the average diameter of the sense coil, l is the core length.

As a rule, the value of d_m does not exceed half of l, hence $\xi_2 \geq 0.85$. In some cases $d_\mathrm{m} \approx l$, and even then the value of ξ_2 does not go below 0.6. Consequently, ξ_2 may be considered to be within the following limits $0.6 < \xi_2 < 1$.

ξ_3 may take different values between 0 and 1, depending upon the degree of the shunt action created by the external metallic parts of the fluxgate which possibly cause short cut turns. In order to make this coefficient be approximately equal to 1, usually cores with thin plates tube cores are used. Such parts are used only in fluxgates with transversal excitation (see Fig. 1.4b), all other parts of the fluxgate are preferably made from non-conductive materials. Assuming $\xi_3 = 1$ we find that ξ, defined via (2.55), will lie between the following limits $0.3 < \xi < 1$. Consequently, the error of calculating G_max with (2.47) may be up to 300 %, if (2.56) and (2.57) are not taken into consideration.

In conclusion, note that many particular cases of estimation of the maximum transformation ratio G_max with (2.47), in a combination with (1.52), (1.54), (2.56) and (2.57), show a good agreement with experimental data. The deviation does not exceed 10–15 %, if $m \leq 10^3$, as was recommended in Sect. 1.4.5, and $H_\mathrm{m} > 3H_\mathrm{S}$. The derived values G_max, as a rule, turn out to be overestimated which may be explained as follows: The real dependence $\mu_\mathrm{d}^*(\omega t)$ differs from the stepwise one in the same way, as the real magnetizing curve $B(H)$ differs from its piecewise linear approximation (see Fig. 2.4 and Fig. 2.6). Only for the rectangular shape of the excitation field the mentioned deviation is minimized.

2.5 Closed Core Fluxgates

2.5.1 Types of Closed Core Fluxgates

Closed core fluxgates find an increasingly wide use due to their advantages over fluxgates with open cores. Among the advantages there are:

- lower noise level and improved stability of the zero level signal;

- low level of the dissipative fields, and as a result, a low level of the misbalance EMF;

- low power consumption due to the reduced volume of the core, made usually from a thin permalloy tape, and also due to the possibility to apply a ferroresonance excitation mode (see Sect. 2.6);

- the possibility of a simple equilibration of ring core fluxgates rotating the ring core with respect to the fixed sense coil, with its center being coincident with the center of the ring;

- the possibility to measure two components of a magnetic field simultaneously (see below).

Ring core fluxgates are the most widely used fluxgates with closed cores (see Fig. 1.2).

As it was noted in Sect. 1.1, the ring fluxgate was in fact invented by H. Aschenbrenner and G. Goubau [51]. More perfect designs of such fluxgates were proposed by Ya. Berkman (see Sect. 1.1.4) and by V. Geiger [56]. D. Gordon [57, 58, 59] improved considerably the characteristics of ring core fluxgates resulting in industrial production level.

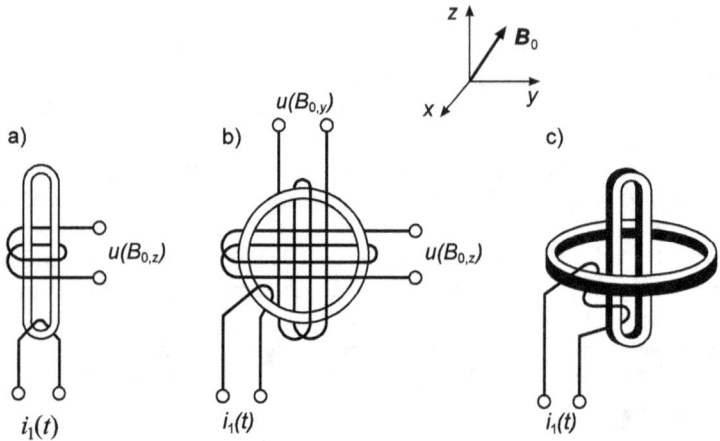

Figure 2.7: Closed core fluxgates: a) single component, b) double component, c) triple component.

In our opinion, the three types of closed core fluxgates shown in Fig. 2.7 are the most promising. They may be used for measurement of one, two and three components of a magnetic field, respectively. The fluxgate illustrated in Fig. 2.7a contains a core in the form of an oblate ellipse, which has two parallel straight line parts in the direction of the main axis. Such a core is also known as a "Racetrack" [54]. It differs from the famous frame core by the more perfect technology of production. While the frame core is stamped, the elliptic core of the considered type is made in the manner used for the production of the ring type core, namely by winding a thin permalloy tape onto a ceramic or metallic base. Being exposed to an external field it is preferably magnetized in the direction of the long axis, that predominates its use in a single component fluxgates.

The ring fluxgate (Fig. 2.7b) is used for a simultaneous measurement of two field components. Due to isotropy of the ring core the corresponding conversion coefficients will be also the same. Such a fluxgate may be used for angular measurements, and in case of three windings it may be applied in systems of synchronous transmission of the angular displacements of some objects in a magnetic field [3, 36].

2.5. CLOSED CORE FLUXGATES

The three component fluxgate (Fig. 2.7c) contains ring and elliptic cores, lying in the mutually orthogonal planes, and a system of three mutually orthogonal measuring coils (not shown in the figure), encompassing the cores such that the axes of the coils and the axes of symmetry intersect in a single point creating a unique magnetic center. Also, two or an even number of elliptic cores [40] may be used in those fluxgates. The fluxgates of such design are rather compact. They provide high measuring precision in both non-homogeneous and homogeneous fields.

2.5.2 Notes for Deriving Formulae

The analysis and estimation is based upon (2.36).

The closed core is normally driven by a circular drive field. For such a field the demagnetizing coefficient $N = 0$. Due to this, the parameters μ^* and μ_d^* in the first determinant of (2.41) ought to be replaced with the material parameters μ and μ_d in correspondence with the discussion in Sect. 1.4. In this case the field $H_1(t)$ turns out to be identical to the internal field $h_1(t)$.

In respect to the measured homogeneous field closed core behave exactly as open cores with the body parameters μ and μ_d. For this reason, there is no need to change parameters μ^* and μ_d^* in the second determinant. In this case, the force of the measured field H_0 differs considerably from the force h_0, which acts inside the core (see Sect. 1.4).

Two approaches for estimating the useful EMF for closed core fluxgates are possible. In the first the force of the measured field H_0 is transformed into the internal force h_0 via (1.28), and then the estimation of the output EMF is in correspondence with the expression that is valid for the magnetic amplifier. In the second approach there is no transformation of H_0 into h_0, but the dependencies $\mu^*[\mu(t)]$ and $\mu^*[\mu_d(t)]$ are directly found in accordance with (1.30) or (1.36). Both approaches are applicable and give the same qualitative results, provided that they are used correctly.

An attempt to estimate the ring fluxgates on the base of the first approach is contained in [57]. Using piecewise linear approximation of the magnetizing curve of the substance $B(h)$ and correct formulas for the estimation of the output EMF of the magnetic amplifier with the longitudinal excitation, its authors derive an incorrect result. The error was due to wrong recalculation of the field H_0 into field h_0, namely

$$h_0 = \frac{1}{1+N\mu_{\text{init.}}} H_0 \, ,$$

where $\mu_{\text{init.}}$ is the initial magnetic permeability.

However, there are no reasons for the use of this parameter. First, because the core remagnetization is performed not along the basic magnetizing curve, but along the mean one (see Sect. 1.3). Second, because in the expression given above the influence on the parameters of the excitation is not considered.

In the frameworks of the parametric theory the correct formula will be:

$$h_0(t) = \frac{1}{1 + N[\mu_d(t) - 1]} H_0 . \tag{2.58}$$

Substituting (2.58) into the expression for the useful EMF of the fluxgate, estimated as a magnetic amplifier, we obtain (considering (1.36) for $H_0 = const \neq 0$):

$$\begin{aligned} u(t) &= -s_\Sigma w_2 \xi \mu_0 \frac{d}{dt}\left(\mu_d[h_1(t)]\, h_0(t)\right) \\ &= -s_\Sigma w_2 \xi \mu_0 H_0 \frac{d}{dt}\left(\mu_d^*[h_1(t)]\right), \end{aligned} \tag{2.59}$$

where s_Σ is the total area of the cross-section of both halves of the closed core.

Note that an expression identical to (2.59) may be obtained directly, i.e. without recalculation of fields with (2.58) from (2.41).

This error was noticed in the West in [53] many years later. In this connection we would like to draw attention to the fact, that long before these publications, the correct formulae for the ring fluxgates estimation were given in works [5, 6]. Using the second approach the authors not only found a good correspondence of the estimated data with experimental ones, but also pointed out the limitation and sometimes inapplicability for an analysis and estimation using the piecewise linear approximations for the magnetization curve $B(h)$.

The method of analysis and estimation of ring fluxgates offered in [5, 6] will be expanded to other types of closed core fluxgates below.

2.5.3 Recommended Method of Estimation

Let us approximate the mean curve of the magnetizing of the core substance $B(h)$ with the arc tangent function

$$B = \mu_0 h + \alpha \arctan(\beta h) , \tag{2.60}$$

where the term $\mu_0 h$ describes so-called paraprocess under $B > B_S$, and B_S is the saturation induction.

For longitudinal excitation we found the following expression for the parameter μ_d:

$$\mu_d = \frac{1}{\mu_0}\frac{dB}{dh} = 1 + \frac{1}{\mu_0}\frac{\alpha\beta}{1 + \beta^2 h^2} . \tag{2.61}$$

In this case the limiting values will be equal to

$$\mu_{d,\,max} = \lim_{h \to 0} \mu_d = 1 + \frac{\alpha\beta}{\mu_0} , \tag{2.62}$$

$$\mu_{d,\,min} = \lim_{h \to \infty} \mu_d = 1 . \tag{2.63}$$

2.5. CLOSED CORE FLUXGATES

It is reasonable to assume that the coefficients α and β in (2.60) are equal to

$$\alpha = \frac{2B_S}{\pi} \quad \text{and} \quad \beta = \frac{\pi}{2h_S}, \qquad (2.64)$$

where h_S is the force of the normalizing field, which under piecewise linear approximation of the curve $B(h)$ has the meaning of the saturation field (Fig. 2.8). In this case for $h \to \infty$

$$B \to (\mu_0 h + B_S)\big|_{B_S \gg \mu_0 h} \approx B_S,$$

$$\mu_{d,\max} \to \left(1 + \frac{1}{\mu_0}\frac{B_S}{h_S}\right) \approx \frac{1}{\mu_0}\frac{B_S}{h_S}$$

This is in good agreement with the physical picture of the phenomena under consideration.

Omitting the first term, equal to one, in (2.61), we find the expression for periodic function $\mu_d[h_1(t)]$ in the mode of sinusoidal excitation field force

$$\mu_d[h_1(t)] = \frac{\mu_{d,\max}}{1 + \left(\frac{\pi^2}{4}\right)\left(\frac{h_m}{h_S}\right)^2 \sin^2(\omega t)}. \qquad (2.65)$$

Substituting this expression into (1.36) and using (2.41) for $H_0 = const \neq 0$, $\beta \approx 0$, and using the estimation method described in [5], we find the coefficient of the fluxgate conversion for the second harmonic of EMF

$$G_2 = 4\omega s_\Sigma w_2 \xi \mu^*_{d,2}(h_m), \qquad (2.66)$$

where

$$\mu^*_{d,2}(h_m) = \frac{\mu_{d,\max}}{1 + (\mu_{d,\max}/m)}\frac{1}{A}\frac{A-1}{A+1}, \qquad (2.67)$$

$$A = \sqrt{1 + \frac{1}{1 + (\mu_{d,\max}/m)}\frac{\pi^2}{4}\left(\frac{h_m}{h_S}\right)^2}; \qquad (2.68)$$

m is the permeability of the shape of the closed core, estimated with (1.53) or (1.54).

The maximal conversion coefficient is achieved under the following relation

$$(h_m)_{opt} = 1.4\sqrt{1 + \frac{\mu_{d,\max}}{m}}\, h_S \qquad (2.69)$$

62 CHAPTER 2. FUNDAMENTALS OF PARAMETRIC THEORY

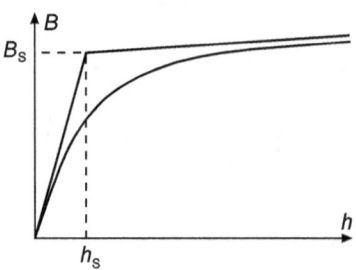

Figure 2.8: The piecewise linear and arc tangent approximations of the mean magnetization curve for closed cores.

Taking into account (1.55) it turns out that

$$G_{2,\,\text{max, sin}} \approx \frac{2.15}{\pi}\omega s_\Sigma w_2 \xi m \,. \qquad (2.70)$$

Using (2.66)–(2.68) it is possible to obtain dependencies $G_2(h_m)$ for various values of m. These dependencies, being normalized with respect to values G_2/G_{max} and h_m/h_S, are shown in Fig. 2.9. It is seen that with decreasing m the maxima of the curves are shifted to the right and higher values of $(h_m)_{\text{opt}}$ are required in order to achieve the maximal transformation ratios for each fluxgate.

In this connection, comparing (2.69) with the corresponding relation (2.50) (for the sinusoidal shape of the drive field), it may be concluded that the qualitative correspondence between them exists only as $m \to \infty$, ie. when the fluxgate starts to play the role of a magnetic amplifier. The shown quantitative correspondence ($1.4 \approx \sqrt{2}$) is evident, in particular, in favor of correctness of the choice of the value of the normalizing field h_S according to (2.64)

As has already been mentioned, the maximal transformation ratio of the fluxgate, independent of the chosen approximation of the mean magnetization curve, may be obtained with a rectangular drive field.

From (2.41) and the other obtained formulae for a closed core fluxgate with rectangular pulse excitation ($\tau_{\text{opt}} = \pi/2$), we obtain

$$G_{2,\,\text{max, rec}} = \tfrac{4}{\pi}\omega s_\Sigma w_2 \xi(\mu^*_{\text{d, max}} - \mu^*_{\text{d, min}}) \,,$$

$$\mu^*_{\text{d, max}} = \left.\frac{\mu_{\text{d, max}} m}{\mu_{\text{d, max}} + m}\right|_{\mu_{\text{d, max}} \gg m} \approx m \,,$$

$$\mu^*_{\text{d, min}} = \left.\frac{\mu_{\text{d, min}}}{1 + (\mu_{\text{d, min}}/m)}\right|_{m \gg \mu_{\text{d, min}}} \approx \mu_{\text{d, min}}$$

2.5. CLOSED CORE FLUXGATES

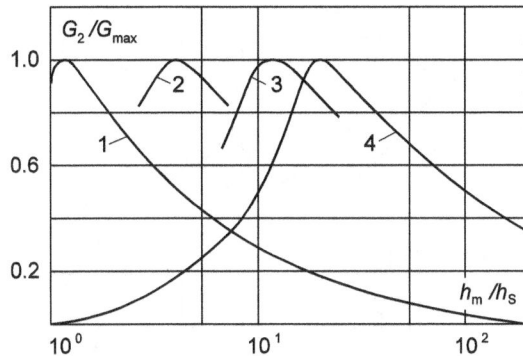

Figure 2.9: The dependencies of transformation ratios of closed core fluxgates on the degree of the relative over excitation. 1: $\mu_{d,\max}/m = 0$ ($m = \infty$); 2:10; 3:100; 4:200.

Under $h_{\max} \gg h_S$ the value $\mu_{d,\min} \to 1$ agrees with (2.63). Hence

$$G_{2,\max,\text{rec}} = \frac{4}{\pi}\omega s_\Sigma w \xi m \ . \qquad (2.71)$$

By comparison of (2.71) and (2.70) we see that with rectangular pulse shape the maximal transformation ratio of the fluxgate is almost two times higher than with a sinusoidal shape of the drive field.

As it was said before (see Sect. 2.4), another advantage of the rectangular pulse excitation of fluxgates is the independence or weak dependence of the transformation ratios G_2 upon the change of the value $H_{\max}$, which, for closed cores, is identical to the value $h_{\max}$.

2.5.4 Extension of the Method to Transversal Excitation

The analysis and estimation method considered above may be extended to fluxgates with transversal excitation, where the cores for the drive field are normally closed (see Fig. 1.4).

An example of the analysis and estimation using the piecewise linear approximation of the mean magnetizing curve of the core substance $B(h)$ and a sinusoidal shape of the drive field $h_1(t) = h_m \sin(\omega t)$ is given below.

The relative dependencies $G_2/G_{\max}$ are plotted in Fig. 2.10 as functions of the ratio h_m/h_S for various values of $\varepsilon = m/\mu_{\max}$. It can be seen that these

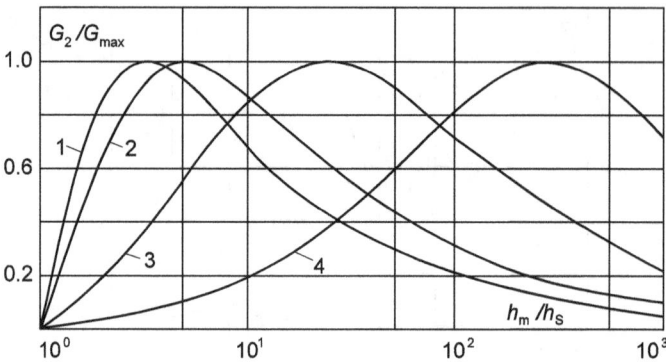

Figure 2.10: The dependencies which characterize the fluxgate operation with transversal excitation. 1: $\mu_{d,\,max}/m = 0.1$; 2: 1; 3: 10; 4: 100.

dependencies are smoother in their character than those shown in Fig. 2.9, however the tendency of the maxima to shift to the right when m decreases, remains here, too.

Fluxgates with transversal excitation are not generally used because of their low transformation ratios (see Sect. 2.3.3), high energy consumption in the excitation circuit, and technological problems when producing tubular cores.

2.6 Modes of Inner Parametric Amplification

The rather simple and physically instructive analysis of the fluxgate performance under load is an example of the effective application of parametric theory.

The works [46, 71] are devoted to the study of fluxgate operation under load, in particular a capacitive load, that leads to an increase of fluxgate transformation ratios.

One practical scheme [48] is depicted in Fig. 2.11. The presence of the capacitive load in the circuit of the sense coil is combined here with a so-called ferroresonance excitation mode of the fluxgate. The conversion coefficient of the fluxgate in this scheme achieved values of the order of 1–10 mV/nT (ie. a hundred times greater than in an off-load run) exhibiting quite satisfactory operation stability at the same time. It is important to understand how the stability of the fluxgate operation is achieved in this scheme and why previously known schemes [46, 71] had not such a stability.

Let us first analyze the reason for the instability in the measuring circuit and then in the excitation circuit.

2.6. MODES OF INNER PARAMETRIC AMPLIFICATION

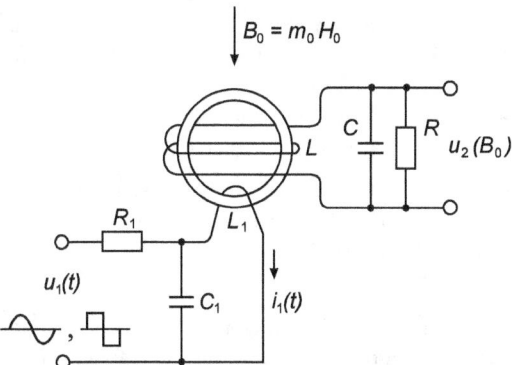

Figure 2.11: The scheme that provides an increase of the transformation ratio of the fluxgate and a good operation stability.

2.6.1 Conditions for Measuring Circuit Stabilization

For time varying inductance of the measuring winding of the fluxgate due to $L \sim \mu_d^*$ it is possible to write a series, analogous to (2.28). Restricting ourselves to the first three terms of the series, we write

$$L(t) = L[\mu_d^*(t)] = L_0 + L_2 \cos(2\omega t) - L_4 \cos(4\omega t) , \qquad (2.72)$$

where L_0, L_2 and L_4 are the constant components and the amplitudes of the second and the fourth harmonics of the inductance $L(t)$, respectively.

After tuning the measuring winding of the fluxgate to the resonance of the frequency $2\omega = 1/\sqrt{L_0 C}$ the current in the circuit can be written in the form

$$i_2(t) = I_2 \sin(2\omega t - \varphi) = i_a \sin(2\omega t) - i_b \cos(2\omega t) ,$$

where φ is the initial phase of the current. The harmonics of the induction voltage we find as follows:

$$U_{2n} = \frac{d}{dt}\left(L(t) i_2(t) \right)$$

For the second harmonic of the voltage we have

$$\begin{aligned} U_2[L_0, L_4(t)] &= 2\omega L_0 i_a \cos(2\omega t) + 2\omega L_0 i_b \sin(2\omega t) \\ &+ \omega L_4 i_a \cos 2(\omega t) - \underline{\omega L_4 i_b} \sin(2\omega t) . \end{aligned} \qquad (2.73)$$

Here the first and second terms express the voltage for resonance inductance L_0, the third and fourth terms express the voltage for inductance $L_4(t)$, and in the last term the underlined product ωL_4 behaves as a *negative* resistance, which just determines the parametric amplification of the signal of the second harmonic.

From this expressions it is not difficult to make the following conclusions:

1. The availability of the fourth harmonic in the structure of $L(t)$ is a necessary condition for ensuring the parametric amplification of the signal of the second harmonic. This absolutely agrees with the theory of parametric resonance [25, 41, 42];

2. The parametric amplification is characterized through the phase selectivity, which can be seen by comparison of the third and fourth terms of (2.73);

3. The parametric amplification arises only where the initial phase of the current is $\varphi > 0$; where $\varphi = 0$ or $\varphi < 0$ parametric amplification does not arise.

In the publication [46] it is shown that the fluxgate transformation ratio is defined by the following expression, where the ferrosonde is working in the parametric amplification mode:

$$G_2^{res} = G_2 \frac{2\omega L_0}{R} \frac{\sqrt{1+K_p^2}}{1-K_p^2},$$

the first term is the fluxgate transformation ratio in the off load mode (which is defined using the formulae in the previous sections), the second term is the quality of the passive circuit, R is the resistance of circuit, and finally the third term is the quality of the active circuit, which is determined using the coefficient of parametric regeneration

$$K_p = \frac{\omega L_4}{R} \;;.$$

In this work it is shown, that the stability of the parametric amplification is ensured where $K_p < 1$.

It is obvious that the stability of the fluxgate parametric amplification is related to the stability of the values L_0 and L_4. As $L_0 \sim \mu_{d,0}^*$ and $L_4 \sim \mu_{d,4}^*$, in order to provide stable parametric amplification, it is necessary, first of all, to be cautious with respect to the stability of the values $\mu_{d,0}^*$ and $\mu_{d,4}^*$.

For the piecewise linear approximation of the mean magnetizing curve $B(H)$ of the cores and for the sinusoidal shape of the drive field, in correspondence with (2.43) we have

$$\mu_{d,0}^* = \frac{2m\theta}{\pi} \qquad (2.74)$$

2.6. MODES OF INNER PARAMETRIC AMPLIFICATION

$$\mu_{d,4}^* = \frac{m}{\pi} \sin(4\theta) .$$

In the given case the saturation angle θ turns out to be independent of the amplitude of the drive field H_m [see (2.48)]. This is why, with a sinusoidal drive field and a change of the field amplitude H_m, it is impossible to provide the required stability of $\mu_{d,0}^*$ and $\mu_{d,4}^*$. The same would be true also for a triangular drive field, considering (2.48).

And it is only for the rectangular pulse drive field that the stability of the values $\mu_{d,0}^*$ and $\mu_{d,4}^*$ may be provided.

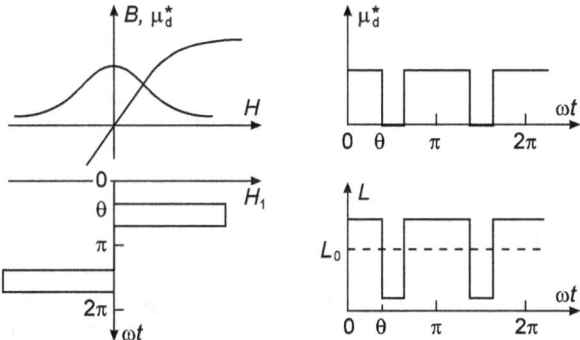

Figure 2.12: Analysis of the fluxgate performance in the inner parametric amplification mode with a rectangular pulse excitation signal.

Corresponding graphical constructions are contained in Fig. 2.12, from where it is seen that independently from the chosen approximation of the mean magnetization curve, the periodic function $\mu_d^*[H_1(t)]$ has a stepwise rectangular shape and is uniquely characterized by the angle θ, which, in this case, does not depend upon maximal field value H_{max} and is defined, according to (2.48), by the pulse duration τ only. Formulas (2.74) remain valid also in this case. It is necessary only to choose the angle θ, i.e. the pulse duration τ, correctly. Under $\theta = \pi/4$ ($\tau = \pi/2$) we derive the conditions that are optimal for the operation of the fluxgate with the output at the second harmonic in the off load mode [see (2.50)].

However, if the sense coil is tuned to resonate with the second harmonic, then no inner parametric amplification and consequently no gain in the transformation ratio of the fluxgate will be obtained. This happens because under $\theta = \pi/4$ the coefficient $\mu_{d,4}^*$, in correspondence with (2.74), turns out to be equal to zero.

For small fluctuations $\pm \Delta\theta$ relative to $\theta = \pi/4$ the parametric amplification of course does take place, yet it will be extremely unstable, because the dependence

$$\sin\left[4(\theta \pm \Delta\theta)\right]\big|_{\theta=\frac{\pi}{4}} = \pm \sin(4\Delta\theta) \approx 4\Delta\theta \qquad (2.75)$$

is considerably steep.

Under $\theta = 3\pi/8$ ($\tau = \pi/4$) we obtain conditions, which are optimal for the operation of the fluxgate with the output at the second harmonic in an on load mode. The coefficient $\mu_{d,2}^*$ in this case becomes slightly less ($\sqrt{2}$ times). However, the coefficient $\mu_{d,4}^*$ achieves its maximal value corresponding to (2.74). Inner parametric amplification takes place and the fluxgate transformation ratio becomes essentially higher.

Under minor fluctuations $\pm \Delta\theta$ relatively to the value $\theta = 3\pi/8$ the parametric amplification remains stable, as

$$\sin\left[4(\theta \pm \Delta\theta)\right]\bigg|_{\theta=3\pi/8} = -\cos(4\Delta\theta) \approx -1 \qquad (2.76)$$

turns out to be very gently sloping. This guarantees the stable operation of the fluxgate.[8]

Consequently, along with the well known methods of achieving stable operation of the fluxgate in the capacitive load mode [46, 71], a rather effective one is to use a rectangular pulse excitation mode provided that, simultaneously, a correct choice of the saturation angle θ is made, the latter should be equal to $3\pi/8$ (for the second harmonic of the output voltage) [40, 68].

The described method to increase the transformation ratio is applicable to fluxgates of any type. Exceptions are only fluxgates with a circularly polarized excitation, as they have $(H_1 dmu^*/dH)_4 = 0$. However, if the polarization is not circular but elliptic, then the coefficient $(H_1 dmu^*/dH)_4 \neq 0$, and the inner parametric amplification in the fluxgates of the described type is also possible.

The sharp increase of the transformation ratio of fluxgates working in the mode of inner parametric amplification allows not only to improve the relation between the second and the odd harmonics of the output voltage (in the off load mode this is the same as the considerable decrease of the EMF of the misbalance), but also between the inphase (useful signal) and the quadrature (noise) components of the voltage of the second harmonic (see Sect. 3.2) as well as between the useful signal of the fluxgate and the noise of the input elements of the electronic components of the device.

2.6.2 Conditions for Excitation Circuit Stabilization

There are certain problems with the operation of a rectangular pulse drive field with low content of even harmonics. This is why in a number of cases one uses a form of current that is close to a rectangular pulse one, which may be implemented, for example, under the ferroresonance drive mode. The scheme of implementation of the ferroresonance drive mode is clear from the Fig. 2.11.

In series with the fluxgate drive coil, playing the role of a nonlinear inductance $L_1(u)$, there is a linear resistor R_1 (sometimes a linear inductance is used), and in

---

[8]The parametric amplification may be achieved also in higher even harmonics, for example, in the fourth harmonic of the output voltage (under $\theta = 3\pi/16$ and $\theta = 7\pi/16$).

2.6. MODES OF INNER PARAMETRIC AMPLIFICATION

parallel to the winding a linear capacitor C_1 is installed. A voltage of a sinusoidal or rectangular shape is fed to the terminals of the circuit. For a sinusoidal voltage a bandpass filter tuned to the frequency of the drive generator may be used instead of the resistor R_1 (or inductance).

Prior to core saturation, the inductance of the drive coil $L_1(t)$ is high and the current $i_1(t)$ flows mainly through the resistor R_1 and the capacitor C_1, charging the latter. When the core is saturated the inductance $L_1(u)$ sharply drops and takes the value L_S, close to the value without a ferromagnetic core.

The charge, accumulated on the capacitor C_1 during the remagnetization of the core from $-B_S$ to $+B_S$, is used to achieve a considerable discharge current $i_1(t)$ during a small time interval. As L_S and r (the ohmic resistance of the drive coil) are small, and the voltage on the capacitor U_C is high, the discharge current $i_1(t)$ may achieve large values. The duration of the saturation state of the core is defined by the values L_S and C_1, as the process of capacitor recharging takes approximately half of the period of the free oscillations ($\sqrt{L_S/C_1} > r$) and ends at the moment when the discharging current achieves zero value. Then the core starts to remagnetize back from $+B_S$ to $-B_S$.

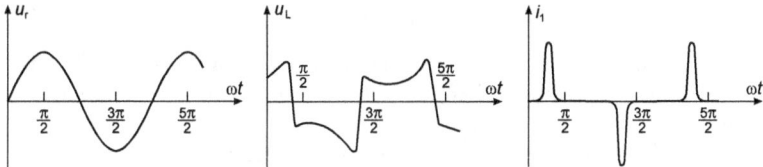

Figure 2.13: The plots of voltages and currents explaining the pulse of the current formation under the ferroresonance mode of the fluxgate excitation.

The plots in Fig. 2.13 explain the essence of the process. It can be seen that the ferroresonance voltage, appearing on the drive coil $u_L(t)$, has the form of rectangles of alternating signs with segments of sinusoids (if the sinusoidal voltage $u_r(t) = U_m \sin(\omega t)$ is fed to the scheme) superposed onto them. The current $i_1(t)$, flowing through the winding, has the form of the alternating sign pulses of a small duration τ with sharp edges. These pulses may be considered to be close to a rectangular shape.

By a variation of the parameters R_1, L_S, C_1 and r, it is possible to achieve conditions, under which the pulse duration τ becomes equal to $\pi/4$ ($\theta = 3\pi/8$). Consequently, a mode will be obtained, which is optimal for the stable parametric amplification of the useful signal in the measuring circuit, ie. for an increase of the transformation ratio of the fluxgate on the second harmonic (see Sect. 2.6.1).

Despite some advantages of the ferroresonance mode of excitation it should be applied with care, especially in cases when the fluxgate is used as an absolute zero-indicator. Cases are known, when the ferroresonance circuit of excitation becomes the source of a false signal in the second harmonic in the measuring circuit. In

70 CHAPTER 2. FUNDAMENTALS OF PARAMETRIC THEORY

these cases, however, the current and the drive voltage (see Fig. 2.13) form did not change. The false signal disappeared when the resistance R_1 was increased (a resistance of 1–3 Ω was used, see Fig. 2.11).

Note in conclusion that the ferroresonance drive mode is best implemented with considerable differences in values of the drive coil inductance as in magnetic amplifiers.

2.7 Magnetic Axis of Fluxgates

2.7.1 The Concept of the "Magnetic Axis"

The fluxgate can be considered as a component converter of the magnetic field. The measured magnetic field $\mathbf{H}_0$ is projected upon an axis inside the fluxgate sensor. This axis is referred to as the magnetic axis and is denoted with the unit vector $\mathbf{i}_M^0$.

The equation for fluxgate conversion at the second harmonic has the following form

$$E_{m,2} = G_2 \mu_0 \mathbf{H}_0 \cdot \mathbf{i}_M^0 + E_{m,2}^f , \qquad (2.77)$$

where $E_{m,2}$ is the amplitude of the second harmonic of the output EMF, G_2 is the conversion coefficient defined by (2.33), and $E_{m,2}^f$ is the amplitude of the second harmonic of a false signal (see Sect. 3.2).

It must be pointed out that the first term of (2.77) may not be considered as the definition of the magnetic axis because the direction of $\mathbf{i}_M^0$ is not defined here at all.

Let us find the dependence of the direction of $\mathbf{i}_M^0$ on the core parameters and the direction of the sense coil's axis.

The process of conversion, occurring in a fluxgate of any type, may be thought of as consisting of three independent operations:

1. the magnetization of the cores in the measured field $\mathbf{H}_0$,

2. the superposition of the drive field $\mathbf{H}_1(t)$ onto the cores with the purpose of affecting the parameter $\hat{\mu}_{\mathrm{dyn}}^*$ and

3. the modulation of the magnetic flux in the cores, and the reading of the information about the modulated flux with the help of a sense coil.

These operations are described in full detail in Sects. 2.2 and 2.3. For a fluxgate operating on the second harmonic we find for the last two operations, respectively,

$$\mathbf{B}_2(\mathbf{H}_0) = \mu_0 \hat{\mu}_{d,2}^*[\mathbf{H}_1(t)] \cdot \mathbf{H}_0 , \qquad (2.78)$$

$$B_{i_w}(H_0) = \mathbf{B}_2(\mathbf{H}_0) \cdot \mathbf{i}_M^0 , \qquad (2.79)$$

2.7. MAGNETIC AXIS OF FLUXGATES

where $\mathbf{B}_2$ is the component of the second harmonic of the induction vector of the core, $\hat{\mu}^*_{d,2}$ being the tensor of the second harmonics of the dynamic permeability of the core (see Sect. 2.3), and B_{i_w} is the projection of vector $\mathbf{B}_2$ onto the axis $\mathbf{i}^0_w$ of the sense coil.

The projection B_{i_w} is the density of the magnetic flux of the second harmonic that penetrates into the measuring coil and causes the EMF $u_2(t)$ in it.

The projection may be defined in two ways. Firstly, on the basis of (2.77) one may write with confidence

$$B_{i_w} = c_2 \mu_0 \mathbf{H}_0 \cdot \mathbf{i}^0_M , \qquad (2.80)$$

where c_2 is the constant coefficient ($c_2 \approx G_2$). Secondly, on the basis of (2.78) and (2.79), taking into consideration the symmetry of the tensor $\hat{\mu}^*_{\text{dyn},2}$, we have

$$B_{i_w} = (\mu_0 \hat{\mu}^*_{\text{dyn},2} \cdot \mathbf{H}_0) \cdot \mathbf{i}^0_w . \qquad (2.81)$$

Resolving (2.80) and (2.81) simultaneously we derive

$$c_2 \mathbf{i}^0_M = \hat{\mu}^*_{\text{dyn},2} \cdot \mathbf{i}^0_w . \qquad (2.82)$$

Expression (2.82) allows one to give the following definition: The magnetic axis of a fluxgate coincides with the direction of the vector resulting from scalar multiplication of the dynamic permeability tensor amplitude of the core with a unit vector, which characterizes the direction of the sense coil axis.

2.7.2 Causes of Destabilization of the Magnetic Axis

The cores of fluxgates are made mainly from a thin permalloy tape, that permits the use of the square matrix of the tensor $\hat{\mu}^*_{\text{dyn},2}$ of the form (2.25), for which $\hat{\mu}_{\text{dyn},x} = \mu^*_{d,2}$ and $\hat{\mu}_{\text{dyn},y} = \mu^*_2$.

As before (see Sect. 2.3.1), we introduce the angles $\alpha \underline{x}^0, \mathbf{H}_0, \beta \mathbf{x}^0, \mathbf{i}^0_w$ and $\gamma \mathbf{x}^0, \mathbf{i}^0_M$ defining the position of the fluxgate magnetic axis $\mathbf{i}^0_M$ (Fig. 2.14) in the chosen coordinate system.

Let us write (2.82) in a matrix form

$$c_2 \begin{vmatrix} \cos\gamma \\ \sin\gamma \end{vmatrix} = \begin{vmatrix} \mu^*_{d,2} & 0 \\ 0 & \mu^*_2 \end{vmatrix} \begin{vmatrix} \cos\beta \\ \sin\beta \end{vmatrix} . \qquad (2.83)$$

Then we derive

$$\tan\gamma = \frac{\mu^*_2}{\mu^*_{d,2}} \tan\beta . \qquad (2.84)$$

We will show, that the angle γ defines the position of the fluxgate magnetic axis. If x and y are the main axes of the excited core (Fig. 2.14), then under $\alpha = (90 + \gamma)°$ we have $\mathbf{H}_0 \perp \mathbf{i}^0_M$, and, consequently,

$$H_0(-\mu^*_{d,2} \sin\gamma \cos\beta + \mu^*_2 \cos\gamma \sin\beta) = 0 . \qquad (2.85)$$

72   CHAPTER 2. FUNDAMENTALS OF PARAMETRIC THEORY

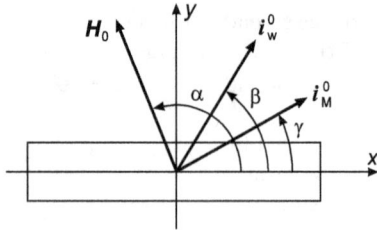

Figure 2.14: The definition of the concept of the magnetic axis of a fluxgate.

Expressing $\sin\gamma$ through (2.84) we see that (2.85) is indeed valid. In the process of fluxgate design one tries to minimize the angle β to be small in order to get the axis $\mathbf{i}_M^0$ of the sense coil to be parallel to the axis $\mathbf{x}^0$, ie. to the vector $\mathbf{H}_1$ of the drive field. Besides, it would not be practical to use cores, for which the condition $\mu_2^*/\mu_{d,2}^* \leq 1$ may be implemented. Consequently, the angle is also small. Under these conditions instead of (2.84) it is possible to write

$$\gamma = \frac{\mu_2^*}{\mu_{d,2}^*}\beta \ . \tag{2.86}$$

By differentiation of (2.86) one may obtain the expression for the increment of the angle γ, which defines the instability of the spatial position of the fluxgate magnetic axis

$$\Delta\gamma = \frac{\mu_2^*}{\mu_{d,2}^*}\Delta\beta + \beta\Delta\left(\frac{\mu_2^*}{\mu_{d,2}^*}\right) \ . \tag{2.87}$$

As the necessity to fix all the parts inside the fluxgate is obvious, it may be assumed that $\beta = const \neq 0$. In this case γ will depend only upon the increment of the ratio $\mu_2^*/\mu_{d,2}^*$, which, in turn, depends upon the drive field $H_1(t)$:

$$\Delta\gamma = \beta\Delta\left(\frac{\mu_2^*[H_1(t)]}{\mu_{d,2}^*[H_1(t)]}\right) \ . \tag{2.88}$$

The difference between the dependencies $\mu_{d,2}^*[H_1(t)]$ and $\mu_2^*[H_1(t)]$ was discussed in Sect. 2.3.2.

Thus, even in case when all the parts inside the fluxgate are properly fixed, the orientation of the magnetic axis turns out to be dependent on the change of the amplitude or shape of the drive field.

Chapter 3

Magnetic Offsets and Noise

3.1 General Statements

Experimentally used macroscopic parameters and characteristics of ferromagnets (see Chaps. 1 and 2) reflect a summarized effect of magnetic internal forces; these parameters and characteristics are also influenced by such external factors as temperature, mechanical stresses, magnetic fields, etc. [8, 11, 13, 14]. Since a ferromagnetic material in its natural state, i.e. being cooled below the Curie point in the absence of an external magnetic field and other forces, becomes demagnetized. The total magnetic moment of the material is

$$M_\Sigma = J_\mathrm{S} \sum_i v_i \cos \theta_i = 0, \tag{3.1}$$

where J_S is the modulus of the spontaneous magnetization vector (the modulus is the same for all domains), v_i the volume of ith domain, and θ_i the angle between the vector $\mathbf{J}_\mathrm{S}$ of the ith domain and any fixed direction in the ferromagnet.

Let us now place the ferromagnetic sample into an external magnetic field of intensity H. The sample begins to be magnetized and a nonzero magnetic moment $\Delta M_\Sigma(H)$ appears in it in the direction of the vector H. In the general case, this moment can be represented in the form

$$\Delta M_\Sigma(H) = J_\mathrm{S} \left(\sum_i \Delta v_i \cos \theta_i + \sum_i \Delta \theta_i v_i \sin \theta_i \right). \tag{3.2}$$

The first sum here reflects the growing volumes of the domains, the vectors $\mathbf{J}_\mathrm{S}$ form acute angles with the vector H. This increase of the volumes of some domains occurs at the expense of decreasing the volumes of other domains, the vectors $\mathbf{J}_\mathrm{S}$ form obtuse angles with the vector H. Because the domain walls are displaced in this process, these changes are called displacement processes.

The second sum in (3.2) reflects changes in the vector $\mathbf{J}_\mathrm{S}$ directions; with increasing H these vectors turn to the vector H direction and finally

74   CHAPTER 3. MAGNETIC OFFSETS AND NOISE

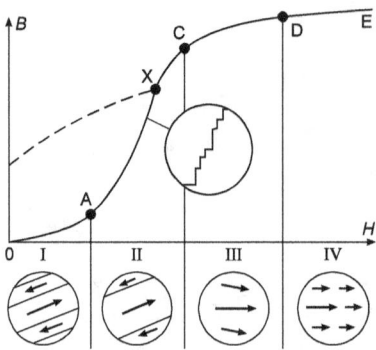

Figure 3.1: Representative regions of the basic curve of ferromagnet magnetization: I – region of reversible movement of domain walls, II – region of irreversible movement of domain walls, III – region of dominant rotation of the spontaneous magnetization vectors and IV – region of paraprocess.

reach this direction. Such variations are generally called rotation processes. The rotation and displacement processes overlap one another and can be reversible or irreversible. The basic magnetization curve has, in general, three characteristic regions (see Fig. 3.1). In the initial region, CD, mainly reversible displacements of domain walls are observed. Therefore after imposing and releasing a weak field H, the domain structure significantly reverts to its initial state. Mobility of the walls at this stage is characterized by the value of the magnetic susceptibility $\kappa_{\text{init.}}$ or permeability $\mu_{\text{init.}}$ (see Sect. 1.3).

In the middle region, AC, we observe irreversible displacements of the domain walls. The steepness of the curve $B(H)$ here is related to an abrupt change in the magnetization (a small part in Fig. 3.1 is shown magnified). This is the Barkhausen effect whose name honors its pioneer (1919). The effect is associated with elastic and irreversible processes overcoming different defects (barriers) by a domain wall, the defects being holes, imperfections of the crystal lattice (dislocations), etc. The higher the magnetoelastic energy of interaction of the walls with the barriers is (see Sect. 3.5), the higher the size of the Barkhausen steps becomes. Mobility of the walls in this part of the magnetization curve is very high and characterized by the permeability $\mu_d \to \mu_{d,\max}$ with increasing field intensity (see Sect. 1.3). When the field is weakened, beginning from any point X on the given part of the curve, hysteresis takes place (the recovery curve is denoted in the figure by broken line).

Observed in the region CD where the field intensity H is still sufficiently great, mainly rotational processes occur. The magnetization here continues to change stepwise, and these changes are mostly related to overcoming the forces of crystallographic and induced (mechanical stresses) magnetic anisotropy. The magnetization changes at this part are dominantly reversible and accompanied by appreciable weakening the hysteresis phenomena. When approaching the point D, we have the limiting value, $\mu_d \to \mu_{d,\min} \approx 1$.

The ferromagnetic sample attains the state of technical saturation at the point D, all the spontaneous magnetization vectors $\mathbf{J}_S$ being oriented in the direction of the vector H. The magnetization has reached its limiting value

$$\lim_{\theta_i \to 0} J(H) = \frac{J_S \sum_i v_i \cos \theta_i}{\sum_i v_i} = J_S \,, \tag{3.3}$$

which is followed from (3.1) providing that the angles θ_i are calculated from the direction of the vector H. The existence of this limiting value is the main factor of nonlinearity of the magnetization curve of ferromagnets, which causes fluxgate units and magnetic amplifiers to work (see Chaps. 1 and 2).

Finally, in the region DE as is clear from (2.3), we have small increases in the magnetization J and the magnetic induction B due to the paraprocess and strengthening of the intensity H. On remagnetization of a ferromagnet along the limiting hysteresis loop (see Sect. 1.3), one in fact deals with similar processes excepting the reversible displacements.

Experiments show that at the output of fluxgate units there is always a spurious signal of the same frequency and, in some cases, the same phase as the legitimate signal.

If the level of this ghost signal remains invariable for a long period, then a displacement or shift of zero is said to take place; if the level of the ghost signal slowly changes, then a drift or slow (infralow-frequency) fluctuations of zero position are said to occur. Finally, if the ghost signal shows itself as a rapid chaotic process possessing a wide continuous spectrum, then we are said to deal with noise in fluxgate units.

The zero offset in a weak constant field H_0 may be found by rotating the fluxgate unit through 180ĉirc about its midpoint. The level of the spurious signal, for example, on the frequency of the second harmonic is determined by (with regard to the phase)

$$E_{m,2}^f = \frac{1}{2n} \left(E_{m,2}^0 + E_{m,2}^{180} \right) i \,, \tag{3.4}$$

where $E_{m,2}^f$ is the amplitude of the spurious signal, $E_{m,2}^0$ and $E_{m,2}^{180}$ are the amplitudes of the total signals, legitimate and spurious respectively before and after rotation, and n is the number of rotations. In the absence of the external field H_0, the spurious signal is immediately revealed. The low frequency fluctuations of the zero position and noise are found out and averaged differently from the the zero offset (see Sect. 3.3).

3.2 Mechanisms of Magnetic Shifts

There are several sources responsible for the appearance of a ghost signal. Some of them occur outside the fluxgate unit. First of all there is the drive-pulse generator which always contains even harmonics, then the frequency doubling circuit feeding the synchronous detector, and, finally, the feedback circuit that is perceiving induction signals from different elements of the electronic circuit of the fluxgate instrument. The methods for suppressing these sources of the ghost signal are considered in Chap. 4. The other sources of ghost signals are inside the fluxgate unit or depend upon specific features of its design and operating mode. Below, we shortly dwell on three characteristic mechanisms responsible for the occurrence of the ghost signal which are internal to the fluxgate unit itself:

1. a mismatch of the fluxgate half cells,

2. the residual magnetization of cores, and

3. a nonuniformity of magnetic fields over the volume of the cores.

3.2.1 Mismatch of the Fluxgate Half Cells

A mismatch of the fluxgate half cells is evident in the presence of a significant level of odd harmonics in the output of the fluxgate unit. This is an unimportant signal if the unit operates in the second regime, in that with the output dependent on the even harmonics (see Sect. 1.2). We are not concerned with the problems of matching the fluxgate units, except for preventing even harmonics from the drive current entering the output signal.

Let us make a simple estimate. Assume that the second harmonic of the generator output amounts 0.01% of the first harmonic, and the coefficient of fluxgate misbalance is $\varepsilon_2 = 10^{-2}$. Then, for a drive field amplitude $H_m = 2000$ A/m, the level of the fluxgate zero offset, $H_{0-\text{offset}} = 2000 \cdot 10^{-6}$ A/m $= 2 \cdot 10^{-3}$ A/m (in field intensity units), or $(B_0)_{0-\text{offset}} = \mu_0 H_{0-\text{offset}} \approx 2.5 \cdot 10^{-9}$ T $= 2.5$ nT (in magnetic induction units). If we take into account that the noise level in modern fluxgate units amounts to $1 \cdot 10^{-11}$ T $= 0.01$ nT or less (see Sect. 3.7), then the estimated zero offset proves to be considerable. Different external actions can affect this level which can be responsible for zero drift.

The question of matching was considered in [77], where it was shown that the achievement of small values of the misbalance coefficient requires careful selection of both cores and windings and by employing active and reactive circuit elements for additional balancing. The given balancing method was used for a long time in practice as applied to bar fluxgate units.

Later on, with the appearance of fluxgates with closed cores made of thin permalloy tape wound on a bobbin, it was found that such instruments have small values of the misbalance coefficient. This fact is explained by two factors:

1. the magnetic properties of both halves of a closed core are more identical to each other than those of two separated cores, because they are made of the same tape, pass the same thermal processing (along with the bobbin) and are not subject to mechanical stresses.

3.2. MECHANISMS OF MAGNETIC SHIFTS

2. due to small scattering fields, the closed cores do not require exact symmetrization of their position relative to the axes of the sense coils.[1]

Another means for improving the balance in ring-shaped fluxgate units is by rotation of the ring core relative to the fixed sense coil (see Sect. 2.5). By using core rotation, the most identical halves can be used. The misbalance coefficient of the ring fluxgate instruments ε_2 can be diminished to the values of $(2 - -5) \cdot 10^{-3}$. The level of the zero offset of these instrument can be reduced correspondingly.

To conclude, we note that for reducing the zero offset, balancing the fluxgate unit is advisable not on the misbalance EMF values for the first and third harmonics, but on the second harmonic, because the conditions for reaching the minimum values of the coefficients ε_1, ε_2, and ε_3 are different. To realize such balancing, one can use an additional low-resistance generator connected in series to the main drive generator and adjusted to a frequency close but unequal to the doubled frequency of the main generator (detuning is 100–200 Hz). The amplitude of the additional generator is set to be not greater than 1% of the working generation current, and the misbalance signal is separated and measured by the spectrum analyzer. The best balance of the fluxgate unit is obtained by minimizing this signal.

3.2.2 Remanent Magnetization

Remanent magnetization of cores is caused by the hysteresis phenomenon (see Sect. 1.3). To exclude the remanent magnetization, reversal magnetization should be made following the limiting hysteresis loop. Since the numerical criterion for attaining the limiting loop in different ferromagnetic materials is different, there are experimental studies [3, 36, 47] of the remanent magnetization as a function of the drive field amplitude (after short-term exposure to strong DC fields).

The remanent magnetization was found to decrease under the condition $H_\mathrm{m} > (H_\mathrm{m})_\mathrm{opt}$ and to be reduced to the minimum at $H_\mathrm{m} \gg (H_\mathrm{m})_\mathrm{opt}$. However, the quantitative relation between the recommended value of H_m and $(H_\mathrm{m})_\mathrm{opt}$ varies depending on the material and shape used. It was also established that, when all things are equal (the same extent of core overexcitation), the residual magnetization is higher in the materials which are characterized by greater values of remanent induction B_R and coercive force H_C (see Fig. 1.6) This magnetization is also higher in cores which have a greater value of the shape permeability m. The cited works point out the role in conserving the remanent magnetization of the surface screening effect of metallic ferromagnets. The next simple formula was

---

[1]The author jointly with V. N. Gorobey and N. A. Chekmareva carried out an experiment in the Research-and production association "D. I. Mendeleyev VNIIM" which was to estimate an effect of asymmetry in the position of closed and opened cores on the misbalance EMF with respect to the same sense coil. They established that at the same deviation of the cores from the coil axis, the misbalance EMF of the closed cores is 5–10 times smaller than that of the open cores.

offered for relating the thickness (or diameter of cross section) of permalloy cores to the admissible frequency f of the drive field [35]:

$$\rho \leq (2-4)f^{-\frac{1}{2}}, \qquad (3.5)$$

where ρ is the core thickness in mm.

When comparing the experimental data on fluxgates with longitudinal and transversal excitation [3, §9] the greater remanent magnetization was found in fluxgate cores with transversal excitation.

A direct relationship between the fluxgate zero drift and the mechanics of decreasing the core magnetization with time was pointed out by Yanus, Drozhzhina, Cheblokov and others [3].

The influence of nonuniformity of magnetic fields over the core volume on the appearance of a ghost signal was found in experimental studies. As was mentioned above, the nonuniformity of the fields within the cores, and over the volume covered by the cores, is always accompanied by the appearance of a signal in quadrature with the true signal at the output. Mikhaylovskiy, Spektor, and Cheblokov [3] showed that the appearance of the quadrature component is associated with the hysteresis-caused phase shift between the magnetic induction and the drive field. If the magnetic flux along the core is nonuniform, for example, because of the core shape (see Fig. 1.12 then the partial (i.e. referred to its individual parts) transformation coefficients and the initial phase of the output EMF also vary along it. This phase change in the second EMF harmonic along the fluxgate core was experimentally confirmed by Mikhaylovskiy and Spektor (see [3]). I. V. Cheblokov studied the quadrature constituent of the output EMF as a function of the field nonuniformity produced by the fluxgate compensating coil. It was demonstrated that the level of the quadrature EMF constituent is dependent on the drive field amplitude. The following actions will lower the level of the ghost signal:

- reducing the length of the sense coil relative to the core,
- using drive windings and compensating coils producing more uniform fields and
- lengthening open cores with the purpose to induce a more uniform magnetic flux in their central part.

The last recommendation looks now absolutely obvious in the light of the advantages of fluxgates with closed cores, in which the changing flux is highly uniform. However, progress was first made from laminated cores to cores approaching ellipsoidal cores in their uniformity and, later on, closed cores [3, 5, 6, 56, 57, 58, 59] were made.

3.2.3 Magnetostriction Effect

In 1969, Weiner [76] drew attention to the seemingly new mechanism of formation of a ghost signal in fluxgate units which is caused by the magnetostriction

3.2. MECHANISMS OF MAGNETIC SHIFTS

phenomenon. Weiner started from the following arguments. The magnetostriction effect, i.e. changing the linear sizes of a ferromagnet under the action of magnetic field, is even. Therefore, the ferromagnetic core being placed in a drive field begins lengthening with the doubled frequency. The core lengthening results in contracting lines of force of the field which cross the turns of the sense coil mounted on the core. In Weiner's opinion, this gives rise to the second harmonic at the fluxgate output. Since the magnetostriction lengthening of the core does not depend on the sign of an instantaneous value of the drive field, Weiner proposed the following formula to estimate the "false" EMF of the second harmonic

$$u_2(t) = -w_2 \big| B_1(t) \big| \delta \frac{dl}{dt},$$

where $B_1(t)$ is the magnetic induction in the core due to the drive field, δ and l are the width and the length of the core respectively, and w_2 is the number of turns in the sense coil. Despite the fact that this formula satisfies the dimension requirements, it is in essence invalid, because the physics of the process is not reflected in it; that is, the formula does not take into account that the direction of the force lines changes with each half period, and an induced additional EMF will thus have an odd-harmonic spectrum, even though the periodic core lengthening develops at the doubled frequency. Within the framework of the parametric theory of fluxgate units, the fallibility of Weiner's arguments is evident. Indeed, with regard to the presentation given in Sects. 1.4 and 2.2, one can take into consideration the separated contribution in the fluxgate output EMF from the temporal variations of the parameters μ_d (or μ) and m. For example, for the longitudinal excitation of closed core we can write

$$\mu_d^*(t) = f\{\mu_d(t), m[l(t)]\}, \tag{3.6}$$

$$\frac{d\mu_d^*}{dt} = \frac{\partial f}{\partial \mu_d} \frac{d\mu_d}{dt} + \frac{\partial f}{\partial m} \frac{dm}{dt}, \tag{3.7}$$

where $\mu_d(t)$ and $m[l(t)]$ describe independent processes caused by the action of the drive field. The case where the second term in (3.7) is zero (because we assumed that $m = const \neq 0$) was considered in Sect. 2.5. The quantities $d\mu_d/dt$ and $d\mu^*/dt$ at $H_0 = 0$ were shown to have an even harmonic spectrum. We noted above that, at the expense of the magnetostriction phenomenon, the derivative dl/dt also has an even harmonic spectrum. It is clear that the same spectrum is characteristic of the derivative dm/dt and, therefore, of the resulting derivative $d\mu_d^*/dt$ determined from formula (3.7).

Thus, taking into account the second term in (3.7) does not give any new features in the fluxgate output EMF spectrum determined in accordance with general expression (2.36): at $H_0 = 0$, the output EMF spectrum contains only the odd harmonics, the even harmonics appearing only in the presence of the measured field H_0. Since the magnetostriction lengthening of core is of the order of 10^{-5}–10^{-6}, the second term in (3.7) serves only as a small correction to the first term and can be neglected in technical calculations. Criticism of Weiner's considerations was presented in [4, 58, 69]. In justification, the positive aspect of Weiner's work [76] should be pointed out as well. It contains experimental data supporting the relationship between the zero offset and noise and the magnetostriction constants of iron-nickel alloy (Permalloy, 79% Ni and 4% Mo) used in fabricating the cores. A similar relationship was experimentally found in the works of other authors independently of Weiner. To interpret this relationship, new and more successful physical models were needed, and they were soon offered.

The considered mechanisms (see Sects. 3.2.1–3.2.2), for the appearance of a ghost signal in fluxgate unit, which act also when the measured field is absent ($H_0 = 0$) or offset by the compensating field ($H_0 + H_\mathrm{C} = 0$), can formally be described in the framework of the parametric approach by using a fictitious field intensity H_0^f. Substituting H_0^f in (2.36) instead of H_0, the even-harmonic spectrum of the ghost signal for a fluxgate unit of any type is found in the same way as the spectrum of the legitimate signal.

In some cases, the field H_0^f has a real physical meaning and can be determined. For example, if the level of the second harmonic of current induced by the drive generator (see Sect. 3.2.1) is known, then the field H_0^f appears to be proportional to the initial amplitude (with regard to the coefficient ε_2) value $H_{\mathrm{m},2}$ of the drive field, because, due to the magnetic rectification [36], the drive field causes the same effect as that produced by the direct field equal H_0^f. In other cases, for example, when a ghost signal is induced by a residual magnetization (see Sect. 3.2.2), the fictitious field can be expressed in terms of the remanent induction B_R, $H_0^\mathrm{f} = B_\mathrm{R}/\mu_0\mu_\mathrm{max}^*$, but such a definition turns out to be useless, because both the values, H_0^f of and B_R, remain unknown. Of even greater difficulty is an interpretation of H_0^f in considering the mechanism of generation of ghost signal related to a nonuniformity of the field in the core volume (Sect.3.2.4). In this case, it would be more convenient not to employ H_0^f, but the notion of asymmetry in the hysteresis loop or a mean curve of the core magnetization.

Figure 3.2 shows the piecewise linear approximation being asymmetric relative to the mean curve of core magnetization on longitudinal excitation. Solid lines correspond to the processes in one fluxgate core, and broken lines to the other (for simplicity, the extent of asymmetry in the magnetization curve of both these cores is assumed to be the same).

It is seen that the periodic function $\mu_\mathrm{d}^*[H_1(t)]$ is now characterized not only by one angle θ (see Fig. 2.4), but by two saturation angles, θ_1 and θ_2, which, for a harmonic drive field, are defined by the expressions

$$\theta_1 = \arcsin\left(\frac{H_\mathrm{S}^+}{H_\mathrm{m}}\right), \quad \theta_2 = \arcsin\left(\frac{H_\mathrm{S}^-}{H_\mathrm{m}}\right), \qquad (3.8a) \qquad (3.8)$$

Contrary to (2.39), the spectral composition of the periodic function $\mu_\mathrm{d}^*[H_1(t)]$ contains both even and odd harmonics. With the help of the known formulae, we find the amplitudes of the even and odd harmonics of the differential permeability

3.2. MECHANISMS OF MAGNETIC SHIFTS

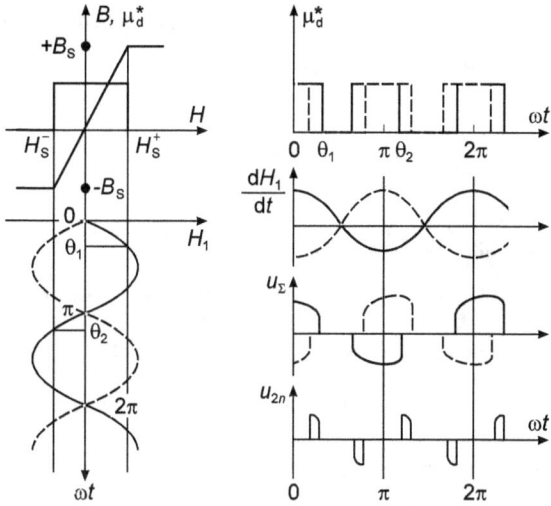

Figure 3.2: To mechanisms of generation of fluxgate ghost signals.

(at $\mu_{d,\max} \gg m$):

$$\mu_{d,2n} = \frac{m}{n\pi}\bigg(\sin(2n\theta_1) + \sin(2n\theta_2)\bigg), \qquad (3.9)$$

$$\mu_{d,2n-1} = \frac{2m}{(2n-1)\pi}\bigg(\cos[(2n-1)\theta_2] - \cos[(2n-1)\theta_1]\bigg). \qquad (3.10)$$

For the symmetrical core magnetization ($\theta_1 = \theta_2 = \theta$), (3.9) converts to (2.43), and (3.10) is zero, which is consistent with the expansion into a series according to (2.39).

Let us apply (3.10) to finding the even harmonics of fluxgate ghost signal in the framework of the autoparametric approach (see Sect. 2.2, [28], and Sects. 4–7 in [4]). At $H_1(t) = H_m \sin(\omega t)$ and taking into account the adopted notations, we have

$$u_{2n}(t) = -\omega s_\Sigma w_2 \xi \mu_0 H_m \frac{\mu^*_{d,2n-1} + \mu^*_{d,2n+1}}{2} \sin(2n\omega t), \qquad (3.11)$$

i.e. the even harmonics of the ghost EMF are represented in terms of the amplitudes of the odd harmonics of the differential core permeability [see (2.39)]. The construction of curves in Fig. 3.2 also indicates the appearance of even harmonics of the ghost signal in the presence of asymmetry in the core magnetization curve.

Let us find the second harmonic of the ghost signal. From (3.10) it follows that

$$\mu_{d,1} = \frac{2m}{\pi}(\cos\theta_2 - \cos\theta_1), \qquad (3.12)$$

$$\mu_{d,3} = \frac{2m}{3\pi}(\cos 3\theta_2 - \cos 3\theta_1)$$

$$= \frac{2m}{3\pi}\left(\frac{4}{3}\cos^3\theta_2 - \cos\theta_2 - \frac{4}{3}\cos^3\theta_1 + \cos\theta_1\right). \qquad (3.13)$$

Summarizing these expressions and substituting the result into (3.11), we obtain

$$u_2(t) = -\frac{4}{3\pi}\omega s_\Sigma w_2 \xi \mu_0 H_m m \left(\cos^3\theta_2 - \cos^3\theta_1\right)\sin(2\omega t). \qquad (3.14)$$

Note that a similar relation for the legitimate signal was derived in [28] for

$$\theta_1 = \arcsin\left(\frac{H_S + H_0}{H_m}\right), \quad \theta_2 = \arcsin\left(\frac{H_S - H_0}{H_m}\right),$$

where H_0 is the intensity of measured field. Rewriting (??),

$$\theta_1 = \arcsin\left(\frac{H_S + \Delta H_S}{H_m}\right), \quad \theta_2 = \arcsin\left(\frac{H_S - \Delta H_S}{H_m}\right), \qquad (3.8b) \quad (3.8)$$

where ΔH_S is the factor of asymmetry of the core magnetization curve, differentiating (3.14) with respect to ΔH_S, we find

$$E_{m,2}^f = G_2 \mu_0 \Delta H_S, \qquad (3.15)$$

where G_2 is the transformation coefficient precisely coinciding with that calculated from (2.49).

Such a coincidence is not accidental and indicates approximately the same observed dependence of legitimate and ghost signals on the drive field amplitude.

The difference between these dependencies is explained by the fact that the measured field H_0 is independent of the drive field, whereas the asymmetry factor ΔH_S inversely depends on the drive field amplitude H_m. Consequently, with increasing H_m, the ghost signal $E_{m,2}^f$ induced, for example, by the core remanent magnetization reduces faster than the legitimate one. Thus, it can be concluded that the parametric theory enables us to describe not only the processes of transformation of legitimate signal, but the processes of generation of ghost signals as well, and to establish the relationship between these processes.

3.3 Magnetic Noise

The Origin of Noise near the Second Harmonic

Noise in fluxgate units and magnetic amplifiers were experimentally found in measuring weak magnetic fields and electric currents. It was clarified that the noise

3.3. MAGNETIC NOISE

sources are concentrated in the ferromagnetic cores themselves [21]. Bittel [52] proposed at least three distinct noise components:

1. the Nyquist noise (also called the thermal or Johnson noise) observed at a constant magnetizing field and temperature,

2. an additional noise observed at a constant magnetizing field and variable temperature, and

3. the remagnetization noise observed at a variable magnetizing field.

The remagnetization noise is dominant, and it is this noise that is mainly called a "magnetic noise", though this reference may also include the second noise component. The magnetic noise is directly linked to the Barkhausen effect (quantized variation of the ferromagnet magnetization). In cyclic remagnetization of a ferromagnet, the step parameters (intensity, duration and initial phase) are not constant, resulting in a magnetic noise. Experiments show that the spectrum of the magnetic noise is periodic (Fig. 3.3). In its spectrogram one can distinguish relatively large uniformregions located between the discrete harmonics of the core excitation frequency, and smaller nonuniform regions adjacent to the harmonics being characterized by a greater spectral noise density.

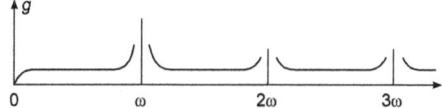

Figure 3.3: A part of spectrogram of magnetic noise observed at the fluxgate output.

The appearance of noise in the regions away from the harmonics can be described as "shot noise" (an analogy to the shot effect in a saturated diode) [21].

The emergence of noise in the regions adjacent to the harmonics does not follow from any normal predictions and is known as the excess noise. Many authors found that the increasing spectral density of the excess noise is governed by the law $1/f^\gamma$, where f is the tuning frequency relative to a harmonic (for a demodulated fluxgate signal it is an observation frequency), and γ is the exponent in the limits 0.5 to 1.5. By analogy with noise of the form $1/f$ in other systems, the excess noise is also named the flicker noise [10, 21].

Possible mechanisms for generation of the excess noise are considered below. Here, we only note that it is the excess noise that, as a rule, determines the resolution threshold of magnetometers.

3.3.1 Modelling Magnetic Noise

Assume that this noise is steady-state and ergodic (this is consistent with experimental data). Then, the noise at the fluxgate output is characterized by the standard deviation

$$\sigma_E = \sqrt{\int_{f_1}^{f_2} g(f)\,\mathrm{d}f}\,, \qquad (3.16)$$

where $g(f)$ is the spectral power density of the noise process, V^2/Hz, f_1 and f_2 are the lower and upper bounds of the studied frequency range, respectively. The noise level at the fluxgate output is described by the following quantity

$$\overline{E}_\mathrm{n} = \kappa \sigma_E E\,, \qquad (3.17)$$

where κ is the coefficient depending on the noise distribution law and a given confidence limit of the estimate. The values of $\overline{E}_\mathrm{n}$ σ_e are measured in Volts. The noise level at the fluxgate input is described by the quantity

$$\overline{B}_\mathrm{n} = \frac{\overline{E}_\mathrm{n}}{G}\,, \qquad (3.18)$$

where G is the fluxgate transformation coefficient (in working on the second harmonic, $G = G_2$, see Chap. 2). The value $\overline{B}_\mathrm{n}$ is measured in Teslas and more conveniently used, since it directly characterizes the fluxgate sensitivity threshold [4, 40]. Also frequently used is the quantity

$$b_\mathrm{n} = \frac{\sqrt{g(f)}}{G}\,, \qquad (3.19)$$

called the spectral density of the amplitude values of the magnetic noise, [T/Hz].

Taking into account (3.16)–(3.19), we find the relation between all the above-mentioned quantities

$$\overline{B}_\mathrm{n} = \frac{\kappa}{G}\sqrt{\int_{f_1}^{f_2} g(f)\,\mathrm{d}f} = \kappa\sqrt{\int_{f_1}^{f_2} b_\mathrm{n}(f)\,\mathrm{d}f}\,. \qquad (3.20)$$

The magnetic noise is characterized by a normal distribution. Therefore, at $\kappa = 2$, the estimated value of $\overline{B}_\mathrm{n}$ reaches the maximum value of the noise EMF (with a confidence of 0.95), which allows the estimate of $\overline{B}_\mathrm{n}$ to be related to the standard peak-to-peak, which are equal to $2\overline{B}_\mathrm{n}$.

The necessity of integral estimates of the magnetic noise is dictated by a nonuniformity of its spectral density. We noted above that the nonuniformity is caused by the presence of the excess noise, the intensity of which increases when approaching discrete lines of the spectrum. However, it is not to be supposed that this noise increases to infinity with decreasing the observation frequency, with the observation time.

3.3. MAGNETIC NOISE

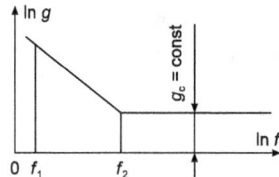

Figure 3.4: Spectral density of magnetic noise power after detection.

Let us make a particular calculation. The spectral density curve of the noise power $g(f)$ of a detected fluxgate signal can be divided into the uniform ($f > f_2$) and nonuniform ($f < f_2$) parts, where f_2 is the upper boundary frequency of the excess noise corresponding to the inflection point on the curve (Fig. 3.4).

For the uniform part, $g(f) = g_S =$, and for the nonuniform part, $g(f) = g_S/f^\gamma$. Taking into account that in view of (3.17) and (3.18), $\overline{B}_n \sim \sigma_E$, we find $\overline{B}_n \sim 2.1\sqrt{g_S}$ for $f_1 = 1\text{--}2$ Hz, and $\overline{B}_n \sim 4\sqrt{g_S}$ for $f_1 = 1\text{--}7$ Hz.

From this calculation it is seen that a considerable increase of the observation time, from several minutes to some months, does not result in the same increase of the excess noise level, and, in practice, this allows the levels to be estimated after comparatively short observations.

An extremely monotonic increase of the excess noise with decreasing the frequency was observed only down to the frequencies of $10^{-2}\text{--}10^{-3}$ Hz [64]. Below 10^{-3} Hz, it would be difficult for the noise to be separated from fluctuations and shifts of the zero position, which have no relation to processes proceeding in the fluxgate cores. Therefore, the reliable lower bound of the excess noise should be taken at the frequency value of the order of 10^{-3} Hz.

As to the upper bound of the excess noise, it is dependent on the amplitude of the drive field and oscillates in the limits from units to hundreds of Hz (see Sect. 3.4). Beyond the upper bound, we observe an usual magnetic noise with uniform spectral density.

Originally, the excess noise near the second EMF harmonic in cyclic remagnetizing ferromagnets was experimentally shown by Goronina [21], and, later on, it was directly observed in fluxgate units by Berkman who suggested that the excess noise in the vicinity of the second harmonic has a physical nature different from that of the usual magnetic noise. However, no particular model of the excess noise was offered.

Weiner [76] made an attempt to account for the excess noise near the even harmonics by the magnetostriction phenomenon. However, as we showed in Sect. 3.2.3, his basic arguments turned out to be invalid. Furthermore, it is unlikely for the excess noise to be attributable to a friction in layers of the Permalloy core, as was supposed in [76], since the excess noise is also observed in the fluxgate devices with single bar cores.

Kolachevskiy et al. [21, 39] assumed that the excess noise near the harmonics of the remagnetization frequency can be due to fluctuations in the magnetic permeability like the flicker noise in granulated resistors is related to resistance fluctuations. However, as seen in Sect. ??, this supposition being isolated from other assumptions, for example from an assumption on the existence at least of the short-lived asymmetry in the hysteresis dynamic loop, cannot explain the noise generation near the even harmonics.

Many authors unify the physical nature for both the usual and the excess magnetic noises and account for the spectrum periodicity with the characteristic ascents near the remagnetization harmonics by the correlation between the Barkhausen jumps occurring in different cycles (see, for example [26]). Following [26] (see pp. 133–135 therein), there are three particular situations (Fig. 3.5).

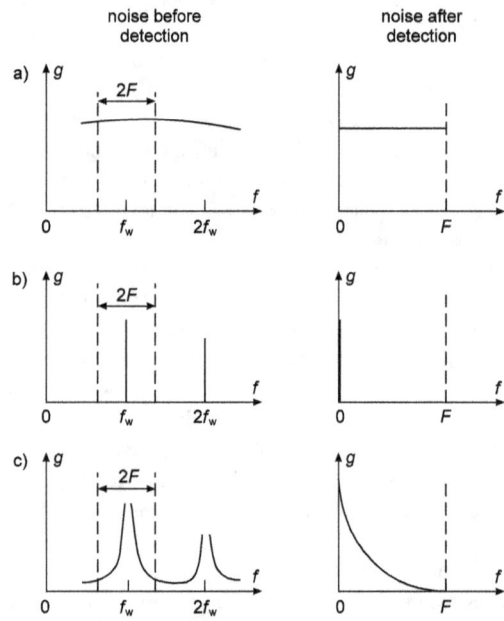

Figure 3.5: Variation of the spectral density of magnetic noise as a function of the extent of correlation between the Barkhausen jumps. (a) The correlation is absent, (b) complete correlation, and (c) incomplete correlation.

- The first of them is the absence of correlations. The Barkhausen jumps appear chaotically and are not repeated from cycle to cycle. The spectral density of the output EMF in the frequency range $2F$ centered at the working frequency f_w is uniform; the low-frequency noise has the spectrum of white noise (Fig. 3.5a).

3.3. MAGNETIC NOISE

- The second situation is the total correlation. The Barkhausen jumps are strictly repeated from cycle to cycle retaining their intensity, duration and initial phase. The output EMF possesses a discrete spectrum; detection produces only the direct component of the process (the shifted zero, Fig. 3.5b).

- The third situation is an incomplete correlation. In every cycle, there are the same jumps, but their parameters (intensity, duration and initial phase) vary from cycle to cycle. The output spectrum tends to be concentrated in the vicinity of the frequency f_w; the density of low-frequency noise after detection decreases with rising frequency (Fig. 3.5c).

It should be noted that the correlation approach was utilized earlier to explain the periodicity of the magnetic noise spectrum. For example, starting from the necessary condition $g(0) = 0$ (the spectral density of noise power is zero), Bunkin [21] showed that the correlation of the Barkhausen jumps separated by an integral number of remagnetization half cycles increase noise on the odd harmonics and decrease noise on the even harmonics. For a long time, this conclusion was used to prove the unsuitability of the correlation approach for describing the noise near the even harmonics. However, as Kolachevskiy [21] noted rightly, Bunkin's result follows from the assumption of conservation of the rigorous symmetry of the hysteresis loop.

The assumption that the loop asymmetry is a source of zero displacements and noise near the second harmonic of the ferromagnet remagnetization frequency was first made by Goronina [21]. It is also the starting point for explaining the ghost fluxgate signal. By using the same assumption Kotlyar proved the applicability of the correlation approach for describing the noise enhancement near the even harmonics of the output EMF. It follows from the relation he derived, that the noise level near the even EMF harmonic is higher, and the Barkhausen correlation is greater for increasing the loop asymmetry in a fluctuation swing.

The cause of the correlation between the Barkhausen jumps can be a slow reconstruction in the defect structure of ferromagnetic material (see Sect. 3.5).

In the framework of the parametric approach (see Chap. 2), it would be natural to relate the appearance of noise to fluctuations of the core magnetic permeability.

As was shown in Sect. 2.3, the differential permeability of cores $\mu_d^*(H_1)$ varies in each half cycle from $\mu_{d,max}^*$ to $\mu_{d,min}^*$ and back. It can be assumed that the value of $\mu_{d,min}^*$ slowly fluctuates. Then, the periodic function $\mu_d^*[H_1(t)]$ also fluctuates.

Let the fluctuation be realized so that the hysteresis dynamic loop in every cycle remains strictly symmetrical. Then, by analogy with (2.27), for the fluctuating spectral part we can write

$$\Delta\mu_d^*[H_1(t)] = \Delta\mu_{d,0}^*(t) + \sum_{n=1}^{\infty} \Delta\mu_{d,2n}^*(t) \cos(2n\omega t + \varphi_{2n}), \quad (3.21)$$

where $\Delta\mu_{d,0}^*(t)$ and $\Delta\mu_{d,2n}^*(t)$ are the fluctuating coefficients of the series (by definition, $\overline{\mu_{d,0}^*(t)} = 0$ and $\overline{\mu_{d,2n}^*(t)} = 0$), and φ_{2n} are the initial phases depending on the coercive force of the dynamic loop.

Inserting (3.21) into the first determinant of (2.36) and using $H_1(t) = H_m \sin(\omega t)$ we find

$$u_n(t) = k\omega s_\Sigma w \xi \mu_0 H_m \cos\beta \left(\frac{2\Delta\mu_{d,0}^*(t) + \mu_{d,2}(t)}{2} \cos(\omega t + \varphi_{2n-1}) \right.$$

$$\left. + \sum_{n=1}^{\infty} \frac{2\Delta\mu_{d,2n}(t) + \mu_{d,2n+2}(t)}{2} \cos[(2n+1)\omega t + \varphi_{2n+1}] \right) \quad (3.22)$$

where k is a coefficient depending on the extent of the spatial correlation of noise in two cores or in two halfs of the same closed core. From expression (3.22) it is seen that the noise in the output EMF caused by fluctuations of permeability is centered only near the odd harmonics of the remagnetization frequency. A noise near the even harmonics of the output EMF is absent, i.e. we obtain a picture similar to that described in [21].

In reality, it would be difficult to imagine a situation in which the permeability slowly fluctuates and the hysteresis loop remains strictly symmetrical. For example, on one-side changing the value of $\mu_{d,\min}^*$, the loop turns out to be asymmetrical in each cycle; the asymmetry being the greater, the higher the rate of increase or decrease of the value of $\mu_{d,\min}$. The symmetry in the hysteresis dynamic loop can be said to exist only at very small values of $d\mu_{d,min}/dt$.

We know from Sect. 3.2 that the presence of short-time asymmetry in the dynamic loop gives rise to odd harmonics in the core permeability spectrum. Series (3.21), as applied to this case, must be supplemented by the corresponding odd part,

$$\Delta\mu_d^*[h_1(t)] = \Delta\mu_{d,0}^*(t) + \sum_{n=1}^{\infty} \Delta\mu_{d,2n}^*(t) \cos(2n\omega t + \varphi_{2n})$$

$$+ \sum_{n=1}^{\infty} \mu_{d,2n-1}^*(t) \sin[(2n-1)\omega t + \varphi_{2n-1}], \quad (3.23)$$

where $\mu_{d,2n-1}(t)$ are the fluctuating amplitudes of the odd harmonics of the permeability (by definition, $\mu_{d,2n}(t) = 0$).

Substituting the odd part of series (3.24) (the expression after the second sum notation) in (2.36), we obtain, by analogy with (3.23), the expression for the noise near the even harmonics of the output EMF

$$u_{n,2n}(t) = k\omega s_\Sigma w_2 \xi \mu_0 H_m \cos\beta \frac{\mu_{d,2n-1}^*(t) + \mu_{d,2n+1}^*(t)}{2} \sin(2n\omega t + \varphi_{2n}). \quad (3.24)$$

This relation shows that the noise near the even harmonics of the output EMF is caused by the fluctuations of the amplitudes of the odd harmonics of permeability, the appearance of which is indicating an asymmetry in the dynamic loop of core

3.3. MAGNETIC NOISE

hysteresis. Bukharov directly related the excess noise near the EMF even harmonics to the fluctuations of the coercive force and of the residual induction of the dynamic loop, providing that the fluctuations are statistically independent, from one half cycle to another. He showed that the relative deviations of 10^{-3}–10^{-4} in the indicated values from the nominal ones are well sufficient to produce a noise level comparable to that observed.

The assumption of the statistical independence of fluctuation in the parameters concerned from one half cycle to another is obviously identical to the assumption on the fluctuating asymmetry in the loop. Therefore, the latter assumption is really the main prerequisite for noise near the even harmonics of the output EMF.

In this connection, we note that the assumption on the fluctuating asymmetry in the dynamic hysteresis loop is not in conflict with the law on the minimum free energy in a ferromagnet (see Sect. 3.1). This law is known to form the basis of the strikingly high symmetry of the limiting loops of ferromagnets (in contrast, for example, to less symmetrical loops in ferroelectrics). However with increasing the sensitivity of recording instruments, the law begins to hold true only "on the average", for a great number of cycles, and thus reveals its statistical nature.

Let us find the relationship of the noise, near the even harmonics of the output EMF, to the external and drive fields.

In order to simplify calculations, we consider an asymmetric mean magnetization curve with the piecewise linear approximation rather than the asymmetric hysteresis loop (see Sect. 3.2 and Fig. 3.2). In doing so, we use the following notations of the core saturation angles characterizing the transformation process

$$\theta_1 = \theta + \frac{\Delta\theta(t)}{2} \quad \text{and} \quad \theta_2 = \theta - \frac{\Delta\theta(t)}{2}, \quad (3.25)$$

where $\theta = \arcsin(H_S/H_m)$ is the average angle of saturation, and $\Delta\theta(t)$ is the fluctuating angle of saturation responsible for an asymmetry in the core magnetization curve at each instant (by definition, $\overline{\Delta\theta(t)} = 0$).

Inserting (3.25) into (3.9) and (3.10), we find, respectively,

$$\mu_{d,2n}^*(t) = \frac{2m}{n\pi} \sin(2n\theta) \cos[n\,\Delta\theta(t)], \quad (3.26)$$

$$\mu_{d,2n-1}^*(t) = \frac{4m}{(2n-1)\pi} \frac{\sin[(2n-1)\theta] \sin[(2n-1)\,\Delta\theta(t)]}{2}. \quad (3.27)$$

(3.27a)

The angle $\Delta\theta$ is usually small. Therefore, the factor $\cos[n\,\Delta\theta(t)]$ in (3.26) is close to unity, and the contribution of the even harmonics $\mu_{d,2n}^*(t)$ to the noise generation near the odd harmonics of the output (sect3.3) is extremely small. Substituting (3.26) in the second determinant of (2.36), we find that, for the same reason, the contribution of the harmonics $\mu_{d,2n}^*(t)$ to the generation of noise (near the even harmonics) proportional to the measured field intensity H_0 is also small.

In any case, this multiplicative component of the noise can be observed in experiments only at greater values of H_0.

Conversely, the contribution of the odd harmonics $\mu_{d,2n-1}^*(t)$ to the noise generation in the output EMF is significant. At small $\Delta\theta$, expression (3.27) takes the form

$$\mu_{d,2n-1}^*(t) \approx \frac{2m}{\pi} \Delta\theta \sin[(2n-1)\theta] , \qquad (3.27b) \qquad (3.27)$$

from which we see that the amplitude fluctuations of the odd harmonics of the permeability are proportional to the fluctuations of the core saturation angles.

Substituting (3.27) into the second determinant of (2.36), we obtain the noise near the even harmonics of the output. In this case, we pay attention to a rather important circumstance: the noise near the even harmonics turns out to be multiplicative relative to the drive field (and to be dependent on the amplitude of this field, see Sect. 3.4) and to be additive with respect to the external field, since the noise exists also in the case when $H_0 = 0$. This conclusion fits the experimental data, which confirm that the basic prerequisite for the noise to emerge near the even harmonics of the output EMF in fluxgate units (and in magnetic amplifiers) is a fluctuating asymmetry of the dynamic loop of ferromagnetic core hysteresis.

3.4 Effects of the Drive Field

Being multiplicative with respect to the drive field, the noise is naturally dependent on the parameters of this field, the amplitude H_m (the maximum value, $H_{1,\max}$, in the case of a wave of unharmonic type), the spatial nonuniformity, $\partial H_1/\partial x$, in the field direction, and the frequency ω.

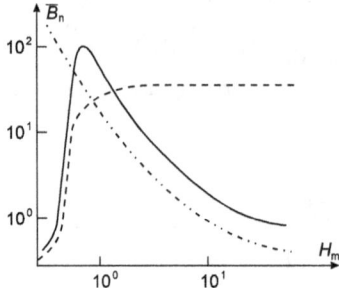

Figure 3.6: The noise level $\overline{B}_n$ versus the amplitude H_m of the drive field.

The solid line in Fig. 3.6 shows the experimentally observed dependence of the noise level $\overline{B}_n$ on the amplitude H_m of the drive field.

The contour of the curve $\overline{B}_n(H_m)$ can be explained, if we represent it as a superposition of the two other curves also shown in Fig. 3.6. One of them (dashed

3.4. EFFECTS OF THE DRIVE FIELD

line) characterizes the increasing number of the Barkhausen jumps with the increasing amplitude H_m, and the other (dash-dot line) characterizes decreasing the fluctuations of the jump parameters. Let us dwell on this in more detail.

At small excitation amplitudes, interactions of domain walls with inhomogeneities and defects of the ferromagnet internal structure are reversible (see Fig. 3.1), the Barkhausen jumps do not occur, and noise in the customary understanding (i.e. as a fast-proceeding random process) is absent. However, a strong zero shift is observed (of the order of 10^{-8}–10^{-6} T), this is due to the presence of a great number of "rigid" regions which do not participate in the remagnetization process and can be viewed as "frozen" fluctuations of the ferromagnet magnetization.

With increasing drive field amplitude, the interaction of domain walls with structural defects becomes irreversible, the Barkhausen jumps appear, and the individual "rigid" regions begin to be remagnetized (though not in every cycle). An obvious representation on these processes is given by the model of disrupted "brittle springs" [21]. An abrupt increase of noise in this field range is associated with increasing the number of the above-mentioned irreversible processes, i.e. the Barkhausen jumps, and with their cycle-to-cycle renewal because of asynchronous remagnetization of the "rigid" regions. Finally, at a certain amplitude of the field, the noise stops increasing. Despite the persistantly increasing number of the Barkhausen jumps, a mechanism suppressing fluctuations of the jump parameters and stabilizing the loop parameters comes into force. The balance of these two trends determines the maximum point on the curve of $\overline{B}_n(H_m)$.

With further increasing drive field amplitude H_m, the dynamic loop tends to its limiting position, and the noise level $\overline{B}_n$ approaches its minimum value.

It is quite important to find a common criterion for estimating the degree of core saturation and, by using this criterion, to evaluate the working amplitude of the drive field at which the noise level does not exceed a preset value. We can agree with the authors of [6], that such a criterion is given by the minimum value of the differential permeability of a material, $\mu_{d,min}$. The value of $\mu_{d,min}$ in fluxgate units is close to $\mu_{d,min}^*$, since

$$\mu_{d,min}^* = \left. \frac{\mu_{d,min}}{1 + \frac{\mu_{d,min} - 1}{m}} \right|_{m \gg \mu_{d,min}} \approx \mu_{d,min}, \qquad (3.28)$$

where m is the permeability of the core shape (see Sect. 2.3).

With the help of the arc tangent approximation of the mean curve of magnetized substance and in view of (2.55), (2.56) and (2.59), we obtain the next expression for the dependence $\mu_{d,min}(H_m)$:

$$\mu_{d,min} = \mu_d\big|_{h_1 = h_m} = 1 + \frac{\mu_{d,max}}{1 + 0.25\pi^2 \left(\frac{h_m}{h_S}\right)^2}, \qquad (3.29)$$

where $\mu_{d,max} = B_S/(\mu_0 h_S) = \mu_d|_{h_1=0}$ is the maximal value of the differential permeability of substance, B_S is the saturation induction, h_1 is the instantaneous value of the drive field intensity (for a closed core, $h_1 \equiv H_1$), and h_S is the intensity of a normalizing field (see Sect. 2.5, Fig. 2.8).

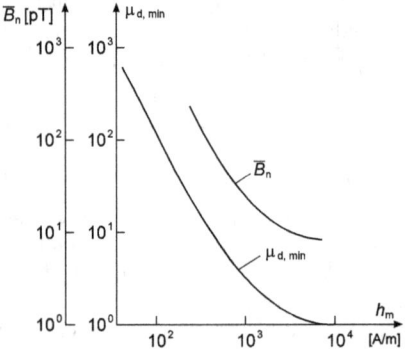

Figure 3.7: The correlation between $\mu_{d,min}(h_m)$ and $\overline{B}_n(h_m)$.

It should be stressed that (3.29) is merely a mathematical model. It is a matter for experiment to test the agreement of this model with the real magnetization curve. Thus, the value of $\mu_{d,min}$ considered as the criterion of material saturation and of a small noise level can be found only from matching the analytical dependence to an experimental curve. We took, as an experimental dependence, the function $\overline{B}_n(h_m)$ to be obtained for a ring fluxgate sample. The selection of the ring fluxgate sample is because of the convenient fitting to the material parameters and of a high reproducibility of these parameters for several samples (as distinct from fluxgate units with cores of any other shape). The core material was an iron-nickel alloy Ni82Mo6Fe[2] with $B_S = 0.67$ T and $\mu_{d,max} \approx 8 \cdot 10^4$. Figure 3.7 gives both the analytical $\mu_{d,min}(h_m)$ [see (3.29)] and experimental relationships $\overline{B}(h_m)$.

We see that these dependencies are strongly correlated. Furthermore, the noise is effectively reduced in the field to 10^3 A/m, then smoothly approaches its limiting value $\overline{B}_{n,min} \approx 10^{-11}$ T (10 pT) at $h_m \geq 2.5^3$ A/m, this value of the field being appropriate to the value of $\mu_{d,min}$ approximately equal to 1.3. Thus, for the material Ni82Mo6Fe the value $\mu_{d,min} \approx 1.3$ is the criterion for reaching a minimum noise in devices using these cores.

The value of h_m required for magnetic amplifiers and fluxgates with closed

---

[2]The following nomenclature is used for characterizing alloys: each component is represented by its chemical symbol which is followed by its mass fraction if known. Iron as the last component adds up to 100%. Example: Ni82Mo6Fe contains 82% Nickel and 6% Molybdenum; the rest is Iron. (+X represents various additives in small amounts.)

3.4. EFFECTS OF THE DRIVE FIELD

cores is calculated with (3.29),

$$h_{\mathrm{m}} = \frac{2}{\pi}\sqrt{\frac{\mu_{\mathrm{d,max}}}{\mu_{\mathrm{d,min}}-1}} h_{\mathrm{S}} = \frac{2}{\pi}\frac{B_{\mathrm{S}}}{\mu_0\sqrt{\mu_{\mathrm{d,max}}(\mu_{\mathrm{d,min}}-1)}}. \quad (3.30)$$

For fluxgates with open cores, one should use

$$H_{\mathrm{m}} = h_{\mathrm{m}} + \frac{B_{\mathrm{S}}}{\mu_0 m} = \frac{B_{\mathrm{S}}}{\mu_0}\left(\frac{2}{\pi}\frac{1}{\sqrt{\mu_{\mathrm{d,max}}(\mu_{\mathrm{d,min}}-1)}} + \frac{1}{m}\right). \quad (3.31)$$

Here, H_{m} is the amplitude of the external field being different for the open core from the amplitude h_{m} of the internal field. Because a drive field is always generated by the drive coil, for closed cores we have $H_{\mathrm{m}} \equiv h_{\mathrm{m}}$. Formula (3.31) thus generalizes (3.30) and really converts to the latter at $m = \infty$.

Using (2.64) and (3.30), we can express the ratio $h_{\mathrm{m}}/(h_{\mathrm{m}})_{\mathrm{opt}}$ in terms of the parameters $\mu_{\mathrm{d,max}}$ and $\mu_{\mathrm{d,min}}$. For $\mu_{\mathrm{d,max}} \gg m$ (usually observed in practice), we have

$$\frac{h_{\mathrm{m}}}{(h_{\mathrm{m}})_{\mathrm{opt}}} \approx \frac{1}{1.4\sqrt{\mu_{\mathrm{d,max}}/m}}\frac{h_{\mathrm{m}}}{h_{\mathrm{S}}} = \frac{2}{1.4\pi}\sqrt{\frac{m}{\mu_{\mathrm{d,min}}-1}}. \quad (3.32)$$

It is seen that the ratio $h_{\mathrm{m}}/(h_{\mathrm{m}})_{\mathrm{opt}}$ turns out to be also dependent on m that is on the permeability of closed core with respect to a measured field. If, now, we specify a value of the ratio $h_{\mathrm{m}}/(h_{\mathrm{m}})_{\mathrm{opt}}$, then different values of $\mu_{\mathrm{d,min}}$ (and thereby different noise levels) will be found for different values of m. This indicates the inadequacy of the "relative overexcitation" criterion which was first pointed out in [6].

In a similar manner, the expression for the ratio $H_{\mathrm{m}}/(H_{\mathrm{m}})_{\mathrm{opt}}$ can be derived. Using (2.50) and (3.31), we find for $\mu_{\mathrm{d,max}} \gg m$:

$$\frac{H_{\mathrm{m}}}{(H_{\mathrm{m}})_{\mathrm{opt}}} \approx 0.707 \frac{0.64 m}{\sqrt{\mu_{\mathrm{d,max}}(\mu_{\mathrm{d,min}}-1)}} + 1. \quad (3.33)$$

Hence, we see again that the "relative overexcitation" criterion is not universal, since now the ratio $H_{\mathrm{m}}/(H_{\mathrm{m}})_{\mathrm{opt}}$ is strongly dependent on the permeability of the open core, m.

By using (3.33), the following problem can be solved as an example: to find such a value of the permeability for a bar core, which will provide a minimum noise level under the condition $H_{\mathrm{m}} = (H_{\mathrm{m}})_{\mathrm{opt}}$. Setting (3.33) equal to unity and substituting the values already known for the alloy Ni82Mo6Fe, $\mu_{\mathrm{d,max}} = 8 \cdot 10^4$ and $\mu_{\mathrm{d,min}} = 1.3$, we obtain $m = 101$. Such value of m, in accordance with (1.52), characterizes a core 10 mm long, 2.4 mm wide and 0.2 mm thick.

This example clearly shows that the criterion of "relative overexcitation" (in the case concerned: overexcitation is absent, $H_{\mathrm{m}}/(H_{\mathrm{m}})_{\mathrm{opt}} = 1$) is in fact impracticable for estimating an extent of core saturation, and the only criterion is the value of $\mu_{d,min}$.

In conclusion, we note that formulae (3.30) and (3.31) are applicable not only to calculation of the amplitudes of the drive field, but also to evaluation of the maximum values of the field of an anharmonic waveform. In such a case, the values h_m and H_m should merely be replaced by the values of h_{max} and H_{max}. Equations (3.32) and (3.33) remain valid only for the harmonic wave form.

According to (3.20), the noise level $\overline{B}_n$ varies inversely with the transformation coefficient G of the fluxgate unit. For this reason, with increasing h_{max} or H_{max}, the noise is reduced most effectively for the rectangular-pulsed drive field, because G does not depend on H_{max} (Fig. 2.5).

Of interest is the following. Which of the core parameters is mostly subject to fluctuations? We assumed above that such a parameter is the quantity $\mu_{d,min}$. Due to the magnetic feedback, the parameter $\mu^*_{d,min}$ is stabilized to an essentially smaller extent in comparison to the parameter $\mu^*_{d,max}$ (see Sect. 1.4) and approaches, as is seen from (3.28), the value of $\mu_{d,min}$.

If, at high core saturation, the value $\mu_{d,min}$ really appeared to be an intensively fluctuating parameter, this would direct the attention of researchers to noise mechanisms related to processes of rotation and nucleation (Sect. 3.5).

Scouten [69] experimentally studied the dependence of noise on the drive field in ring core designs with two sense coils of different diameter. He found that with increasing the field amplitude, the noise falls more quickly in the coil having a greater diameter. Later on, similar results were obtained by other authors [55].

Scouten attributed this dependence to the existence of "noise dipoles" of sufficiently small size which are activated only at the moments of high core saturation, when $\mu_{d,min} \to 1$, but they cannot be observed at other times, when $\mu_{d,min} \gg 1$, because of their strong "magnetic coupling" with the core.

The reference to "magnetic coupling" should be understood as an indication of the existance of a spatial noise correlation.

It should be noted that still earlier Grachev [16] and then Storm and Heiden [74] revealed the basic regularities in the spatial correlation of magnetic noise. The correlation over the core length is manifested in the noise coherence which can be easily found from a significant reduction of noise in two narrow coils which are connected contrarily on the core. They may be located, for instance, at a distance called the correlation interval, where the interval of spatial correlation is defined as the distance $l = l_c$ between two arbitrary points on the core's longitudinal axis such that

$$r_c = e^{-l/l_c} \approx 0.37 \ . \tag{3.34}$$

The quantities l_c and r_c are determined experimentally using movable coils on the core and measuring

a) the mean square of the residual noise EMF,

$$u^2_{N(1-2)} = \overline{(u_1 - u_2)^2} = \overline{u_1^2} + \overline{u_2^2} - 2\overline{u_1 u_2} \ , \tag{3.35}$$

3.4. EFFECTS OF THE DRIVE FIELD

and

b) the mean square of the noise EMF in each coil,

$$\overline{u_1^2} = \overline{u_2^2} = \overline{u}_{N(1,2)} \ . \tag{3.36}$$

Then, for every l_c, the correlation coefficient r_c is calculated using

$$r_c = 1 - \frac{\overline{u}^2_{N(1-2)}}{2\,\overline{u}^2_{N(1,2)}} \ . \tag{3.37}$$

The correlation interval was found to expand with improving the magnetic material properties, especially after material annealing, and the interval length for the permalloy materials is within the limits of 20–40 mm. The best conditions for the noise correlation over the length are provided by ring cores [5, 12, 36]. This explains why ring core designs possess the lowest noise level, all other properties being equal (the same material and the same extent of magnetic saturation) [58, 59, 64].

Returning to the experiment of Scouten [69], we note that the sizes of the "noise dipoles" seem to be small at any value of the excitation field. However, in weak fields, when the value of the minimum permeability is still great, $\mu_{d,\min} \gg 1$, their small-scale structure cannot be revealed because of a strong spatial noise correlation. The correlation becomes weaker only in the strong fields, when $\mu_{d,\min} \to 1$.

This result does not allow the conclusion to be made that the residual noise observed at a high core saturation will necessarily be related to the rotation and nucleation processes. This can be judged from the data obtained for units excited by a rotating field. On the usual excitation of cores, deterioration of noise coherence can occur at any part of the loop including the part in which the displacement processes dominate, and in such a case the residual noise will be related just to these processes.

An effective reduction of noise with increasing the amplitude of the drive field is achieved at a high uniformity of this field inside cores. When open cores are used, a highly uniform drive field is produced when their form approaches an ellipsoid, for example, in laminated cores made in the form of "pointers" [3]. Experiments show that, in designs with such cores, the noise is really reduced. Nevertheless, the efficient means for enhancing the uniformity of the internal drive field is in using closed cores. The following experiment shows to what extent the nonuniformity of the drive field can affect the noise level even for closed cores. The experiment was conducted with a ring fluxgate unit. The core along with the drive winding could be revolved within the fixed sense coil. In doing so, the core revolution plane was made coincident with the coil axis so that the core position was uniquely characterized by the angle α between the coil axis and the line passing through the ring center and an arbitrary point A (Fig. 3.8). The working current $i_1(t)$ was conducted to the drive coil, and the noise EMF $u_n(t)$ was taken from the sense coil.

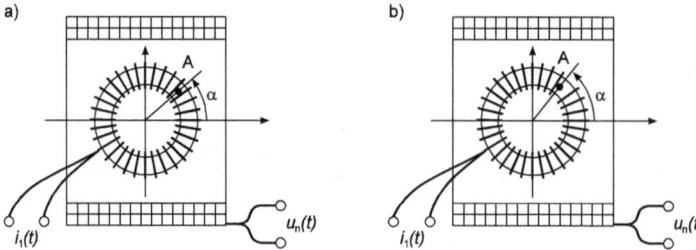

Figure 3.8: Estimating the influence of uniformity of the drive field in ring core on the fluxgate noise level at (a) uniform and (b) nonuniform distribution of drive coil turns.

At first (Fig. 3.8a), inside the sense coil, a homogeneous ring core with the drive coil equally distributed along its circumference was placed. In this case, an appreciable variation in the noise level depending on the revolution angle α was not observed. Then (Fig. 3.8b) several turns (5 from 130) were removed from a small part of the core, near the point A, and this resulted in an essential change of the noise level on revolving the core through the angle α. The noise reached its maximum (i.e. it increased 5–8 times) at α equal to 90° and 270°, and its minimum (an increase was only 1.5 times) at α equal to 0° and 180°. Similar noise variations were also observed by shorting the same turns.

Nonuniformity of the internal drive field also depends upon the inhomogeneity of the core itself. In wound cores, a strong inhomogeneity occurs at the joints of the permalloy tape. The greatest noise level (10–50 times!) is present in fluxgate designs with a core containing only one turn of permalloy tape. In this case, if we consider the tape connection point to be located at point A, the dependence of the noise level on the core rotation angle is similar to that described above. The noise can be lowered only if a second turn shunting the joint is used. Upon adding a third turn, the noise also decreases, but it is essentially independent of superimposing more turns. Three turns seem to be the minimum number which is acceptable for fabricating cores of low-noising fluxgate units. As was noted in Sect.3.2.3, the nonuniformity of the internal field always produces phase shifts. These are responsible for violating the coherence of noise processes in the two halves of a closed core. Thus, the nonuniformity of the internal drive field is, first of all, the factor retarding the spatial noise correlation over the core length.

From experimental data we can conclude that there are many general features between the mechanisms of origination of noise and ghost signal in the form of shifts or drifting of the zero position. First, both noise and zero drifting are reduced with increasing the drive field amplitude. Second, the so-called "rigid" regions responsible for the greater noise level at small amplitudes are nothing more nor less than a time drift of the remanent core magnetization. Third, both the noise and the zero drift are diminished with enhancing the uniformity of the internal drive fields. All this is evidence of a link beetween the noise and zero drifts, or at least, of a link between the usual and excess noise.

3.5 Noise, Magnetostriction, Magnetic Anisotropy

Let us consider the relationship between noise and magnetostriction and magnetic anisotropy. We noted that the noise near the even harmonics of the output of fluxgate units and magnetic amplifiers is due to the destruction of the symmetry in the hysteresis loop of ferromagnetic cores, and the fluctuating character of these violations is caused by, on average (i.e. for the great number of cycles), the law of the minimum free energy in ferromagnetic materials.

It is convenient for a ferromagnet to be represented as a system with negative feedback, caused mainly by long-range magnetic forces rather than by exchange ones. The magnetostriction forces result in the formation of a domain structure (see Sect. 3.1) and change somewhat after each structural reconstruction. The domain structure is reconstructed in a solid with all its associated defects of the internal structure (inclusions, hollows, microstresses, dislocations, etc.). These imperfections are not at fixed locations in the material. The defects change their orientation and/or position due to their elastic interaction with moving domain walls, the interaction being accompanied by irreversible phenomena (Barkhausen jumps) and consumption of magnetic energy. Consequently, concurrent with the reconstruction of domain structure, the defect structure is also rebuilt, which is to be taken into account when the system with feedback becomes stable,i.e. to the state with a symmetrical hysteresis loop.

The difference in the duration for rebuilding the domain and defect structure deserve attention. The domain structure is in fact "erased" in every cycle of remagnetization. The defect structure alters during many cycles, because changes in the orientation and location of defects obey the "slow" laws of plastic deformation. This postulates a "many-scale" character of the processes proceeding in periodically remagnetized ferromagnets and, eventually, being responsible for the appearance of the excess noise.

Of course, it is tempting to explain the appearance of the excess noise in ferromagnets in the context of the comprehensive (extended to physical biological and other systems) "scaleless" model of the excess noise. However, the "multi-scale" character of fluctuating processes in a number of systems, including periodically remagnetized ferromagnets, is so evident that the proponents of the "scaleless" excess noise themselves also do not reject this character [10].

The parameters of magnetostriction and magnetic anisotropy serve as a measure of the interaction between the ferromagnetic domain and defect structures. For example, the interaction between a domain wall and a dislocation is described by the Pich-Keller expression

$$\mathbf{p} = -\int d\mathbf{l} \times (\hat{\sigma}^{M} \cdot \mathbf{b}), \qquad (3.38)$$

where

$$\mathbf{p}$$

is the force of the interaction, $d\mathbf{l}$ is the length of element 1, $\hat{\sigma}^{M}$ is the tensor of internal magnetostriction stresses caused by altering the magnetization directions in the wall itself, and

$$\mathbf{b}$$

is the Burgers vector. Reeder calculated the internal stresses for the flat walls occurring in iron, nickel and cobalt. It was shown that in a system of coordinates with the axis z normal to the wall, only three components of the tensor $\hat{\sigma}^{M}$ can be used. For a linear dislocation arranged parallel with the plane of the wall, it follows from (3.38) that

$$p_z = \sigma_{11}^{M} l_y b_x - \sigma_{22}^{M} l_x b_y + \sigma_{12}^{M}(l_y b_y - l_x b_x), \qquad (3.39)$$

where l_i and b_i are the components of the vectors l and b.

Minimization of the force p_z requires decreasing the values of the components σ_{11}^{M}, σ_{22}^{M} and σ_{12}^{M} of the tensor $\hat{\sigma}^{M}$. The problem reduces to decreasing the associated magnetostriction constants λ_{100} and λ_{111}.

The constants λ_{100} and λ_{111}, of iron-nickel alloys (which are mostly used in fluxgate units) are known to pass through zero provided the nickel content is 80–83% [9, 13, 29, 33]. Hence we derive an important conclusion: alloys with such low nickel content are characterized by a low intensity of the Barkhausen jumps and therefore by the lowest noise level (see Fig. 3.6).

This conclusion is consistent with the experimental data obtained earlier [3, 36, 76] and with the recent observations of new iron-nickel alloys.

Studies by Afanassiev et al. used iron-nickel alloys (developed in the Institute of precise alloys "I.P. Bardeen CSRI of Ferrous Metallurgy under the guidance of V. V. Sosnin) characterized by a low value of the magnetostriction constants λ_{100} and λ_{111}. In this case, the comparison was made with respect to the value of the saturation magnetostriction λ_S, related to the constants λ_{100} and λ_{111} by the simplified Akulov's formula (see p. 518 in [8])

$$\lambda_S = \frac{2\lambda_{100} + 3\lambda_{111}}{5}. \qquad (3.40)$$

Four core types were used in the study. The noise level was measured after placing each fluxgate into a ferromagnetic screen, the value averaged over three samples of each core type made from each alloy being taken as the measured value of $\overline{B}_n$.

3.5. NOISE, MAGNETOSTRICTION, MAGNETIC ANISOTROPY

Alloy	Magnetic parameters				Noise level of the fluxgate $\overline{B}_n$, [nT] (for the frequencies 0.001–1 Hz)			
	λ_S [10^{-6}]	$\mu_{init.}$ [10^3]	H_C [A/m]	B_S [T]	No. 1	No. 2	No. 3	No. 4
Ni79Mo3Fe	2.0	14	3.20	0.85	—	0.280	—	—
Ni79Mo-3Fe	2.0	20	2.40	0.73	0.82	0.220	0.17	—
Ni80Cr3Cu-0.2Fe	—	22	2.40	0.63	—	0.140	—	—
Ni83V5Fe	0.5	58	1.60	0.60	0.40	0.065	—	0.028
Ni81Mo	0.5	66	0.88	0.50	0.34	0.060	0.06	0.015
Ni82Mo6Fe	$-(0.5..1.5)$	64	1.35	0.67	0.27	0.050	0.04	0.012
Ni84Fe	-14	—	—	—	2.00	—	—	—

Table 3.1 Noise levels of different iron-nickel alloys in dependence of their magnetic properties. (Burning of the alloy Ni81Mo was performed at 800°C, cooling with a gradient of 200°C/h and with 400°C/h; the alloy Ni82Mo6Fe was heated to 1000°C and cooled down with a gradient of 100°C/h in the range from 600 to 350°C.) No. 1 – rod cores seized 10×1.2×0.05 mm, drive frequency $f_1 = 25$ kHz; No. 2 – rod cores seized 130×2.8×0.1 mm, drive frequency $f_1 = 1$ kHz; No. 3 – stamping frames (outer dimensions 20×6 mm, inner dimensions 18×4 mm thickness 0.05 mm), drive frequency $f_1 = 25$ kHz; No. 4 – wound rings (diameter 13.2 mm, width 1.5 mm, thickness 0.12 mm), drive frequency $f_1 = 12.5$ kHz.

The results of this study are listed in table 3.1 giving not only the values of the saturation magnetostriction λ_S, but also the values of the initial permeability $\mu_{init.}$, the coercive force H_C and the saturation induction B_S.

It is seen that the noise level $\overline{B}_n$ correlates with the value of $|\lambda_S|$. However, as is illustrated by the graphs of the dependencies $\overline{B}_n(\lambda_S)$ in Fig. 3.9, the $\overline{B}_n$'s have different slopes at the positive and negative values of λ_S respectively.

It is also seen that the noise $\overline{B}_n$ decreases with increasing initial permeability $\mu_{init.}$ and rises with enhancing the coercive force H_C. These dependencies can be interpreted as follows: if the main contribution to the noise power is really given by inhibiting domain walls at the defects, then the spectral density g of the noise EMF must be proportional to the volume density of the defects ρ, i.e. the proportionality must be observed

$$\overline{B}_n \sim \rho^{1/2} . \tag{3.41}$$

On the other hand, from the theory of plastic materials applied to ferromagnets it follows that

$$H_C \sim \mu_{init.}^{-1} \sim \rho^{1/2} , \tag{3.42}$$

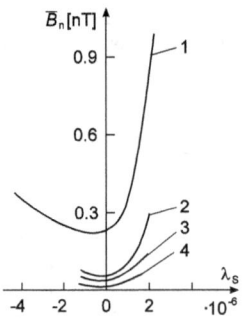

Figure 3.9
Relation between the noise level $\overline{B}_n$ and the alloy saturation magnetostriction λ_S from tests of different fluxgate instruments. 1–4 fluxgate units No. 1, 2, 3, and 4 respectively (the core data are listed in table 3.1).

and therefore the fact that $\overline{B}_n$ is related to the values of $\mu_{init.}$ and H_C turns out to be quite natural. The various absolute values of $\overline{B}_n$ listed in table 3.1 for every alloy and referred to the corresponding types of cores are explained by the facts given in Sect. 3.4:

- the extent of the core saturation (different values of $\mu_{d,min}$) was variable, and

- there was a different nonuniformity of the drive field and a different coherence of noise processes along the cores.

The highest noise level was observed in short bar cores, and the lowest in wound ring cores, which is not in conflict with the conclusions in Sect. 3.4.

The thickness of domain walls is determined by a balance of the exchange and magnetic anisotropy forces; the former tend to increase the wall thickness, and the latter ones tend to diminish it [8, 13]. Therefore, the effects of crystallographic and induced anisotropy should also be taken into account in considering the interaction of domain walls with defects.

Because magnetic anisotropy acts not only on displacement processes, but also on rotation processes, when the Barkhausen jumps also appear, reduction of this effect must result in a reduction of the noise level. The magnetic-elastic anisotropy induced by the action of external mechanical stresses σ is proportional to the magnetostriction constants λ_{100} and λ_{111}, and at a weak magnetostriction anisotropy (i.e. at $\lambda_{100} \approx \lambda_{111} \approx \lambda_S$), it is determined by the product $\lambda_S \sigma$. Therefore, the magnetic-elastic anisotropy is reduced with decreasing λ_S due to using the alloys containing $80 - -83\ \%Ni$.

It is more complicated for the crystallographic anisotropy to be minimized. The magnetostriction constants and magnetic anisotropy constants (usually the constant K_1) for iron-nickel alloys are known to pass through zero at different contents of nickel [29, 33]. Therefore with intent of attaining values of λ_S close to zero by using alloys with nickel content of $80 - -83\%$, we are minimizing the constant of anisotropy K_1 at the expense of applying particular regimes of thermal processing of materials.

A guide to magnetic anisotropy of a material is the highly-rectangular form of its hysteresis loop (ratio B_R/B_S) and low initial permeability $\mu_{init.}$. It is appropriate to follow the course of the dependencies of these quantities versus thermal processing regime, especially for alloys with the low values of λ_S (see table3.1).

These dependencies for the alloy Ni83V5Fe (alloyed with vanadium) were obtained by Shtsherbakovin et al. The ratio B_R/B_S was found to decrease from $0.85\ to\ 0.7$ with increasing the annealing temperature (from 600 to $1100°C$), and, additionally, the permeability $\mu_{init.}$ increases from $2 \cdot 10^4$ to $8 \cdot 10^4$, and the maximum, $\mu_{max} = 8 \cdot 10^4$ is reached at the cooling rate of $100°C$ per hour (in the temperature range $600\ to\ 350°C$). In this case, the observed noise minimum is coincident with the point of maximum, $\mu_{max} = 8 \cdot 10^4$. The authors correctly interpreted the result by relating the noise minimum to obtaining the minimum value of the anisotropy constant K_1.

It is of interest to investigate the extent to which the two factors, magnetostriction and crystallographic anisotropy, separately affect the noise in a single material rather than in different materials.

For this reason the following experiment was performed: the used material was a rolled highly-deformed ($\varepsilon = 97.5\%$) tape 0.1 mm thick, which was made from the iron-nickel alloy Ni82Mo6Fe.[3] Samples of different form were made from tape cut at different angles w.r.t. the direction of tape rolling. Recrystallization annealing of the samples was conducted at temperature of $1100°C$ followed by fast (at the rate of $2 \cdot 10^3 (°C/h)$ for the first group of samples) and slow ($100°C/h$ for the second group of samples) cooling in the temperature interval $600\ to\ 300°C$. The samples from the first group possessed clearly defined magnetic anisotropy (they had the rectangular hysteresis loop in the tape rolling direction), and the samples from the second group proved to be more or less isotropic.

Figure 3.10 shows the experimental results in the form of dependencies

---

[3]The parameters and properties of this alloy are described in the catalog "New precise alloys" (Chermetinformatsiya, Moscow, 1979).

of the noise level $\overline{B}_n$, the saturation magnetostriction constants λ_S, the coefficient of the rectangularity of the loop B_R/B_S, and the initial permeability $\mu_{init.}$ on the angle α formed between the vector of drive field and the tape rolling direction.

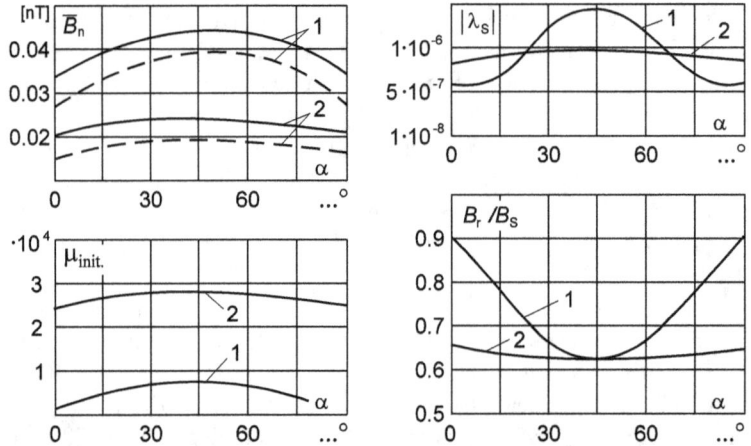

Figure 3.10: The relation of the noise w.r.t the saturation magnetostriction and magnetic anisotropy. 1 – Samples from the first group, 2 – samples from the second group (solid and dashed lines show the curves for different-type fluxgate units).

It is seen (curve 1 in Fig. 3.10) that increasing λ_S by almost one order (at $\alpha = 45°$), the noise level increases only 1.5 times, i.e. it is not as strongly dependent as is expected from Fig. 3.9. This small increase of the noise can only be due to large (at $\alpha = 45°$) reductions in the magnetic anisotropy, an index which represents the coefficient of rectangular form, B_R/B_S. In other words, the two opposing factors (λ_S increases, and B_R/B_S decreases) affected the noise balance in the experiment concerned.

In testing the samples from the second group (curves 2 in Fig. 3.10), all the dependencies become gently sloping. Compared to the dependencies for the samples from the first group, the noise level $\overline{B}_n$ is halved, the saturation magnetostriction takes an intermediate value, the ratio B_R/B_S is close to the minimum value, and the initial permeability $\mu_{init.}$ increases.

Hence, it can be concluded that the noise is indeed related to two factors, magnetostriction and magnetic anisotropy. Furthermore, the lowest noise level is achieved when both these factors are minimized simultaneously (samples of the second group, curves 2 in Fig. 3.10) rather than when one of the factor is minimized at the expense of the other one (samples of the first group, curves 1 in Fig. 3.10).

Thus, the best materials for low noise in fluxgates are alloys characterized not only by a low saturation magnetostriction λ_S, but also by an isotropic magnetic structure having a nonrectangular hysteresis loop ($B_R/B_S \to 0.5$) and increased initial permeability ($\mu_{init.} \to \mu_{max}$), which is in agreement with the former results. Some of these materials are the above-mentioned iron-nickel alloys Ni81Mo, Ni82Mo6Fe and Ni83V5Fe. Amorphous materials are also promising alloys [72].

Let us return to the interaction between domain and defect structures in cyclic remagnetization of ferromagnets.

Since action is equal reaction, from (3.38) it follows that not only does a dislocation act on a domain wall with the force p retarding wall motion, but the wall also acts on the dislocation with the same force of the opposite sign. However, the dislocation responds to this action more slowly. Dislocations possess a low mobility governed by the laws of plastic deformation. The speed of a dislocation (at small stresses) is 10^{-9} m/s or less. Conversely, domain walls are very mobile. This is why a "recurrence" of the interaction of walls with the same dislocations during hundreds and thousands of remagnetization cycles is observed. This is the physical prerequisite for the Barkhausen jumps to be established and so is the cause of the excess noise appearing near the discrete harmonics of the remagnetization frequency (see Figs. 3.3 and 3.5).

Slow changes in the defect structure are also responsible for the fluctuations of asymmetry of the hysteresis loop. The appearance of the asymmetric microhysteresis loops in ferromagnets can be considered as a "barrier effect". A new domain is sometimes produced by passing a domain wall through a dislocation, its appearance being dependent on the direction of motion of the wall; if the domain evolved from a direct motion of the wall, then the back motion of the wall will "erase" this domain. In spite of a rapid adaptation of the domain structure to the dislocation one, the fact of rebuilding of the latter, which is characterized by a "slow time", creates situations in which the total barrier effect at the ascending branch of the loop dominates over that at the descending branch of the loop. This just leads to asymmetry of the loop expressed in the form of different values of the positive and negative coercive forces.

It is obvious that defects affect not only the displacement processes, but also they affect the process of rotation and nucleation. By inducing a local anisotropy, these defects hinder the synchronous remagnetization (i.e. with the frequency of drive field) of the "rigid" regions. The characteristic time delay in this case is related to the kinetics of rebuilding the defect (in particular, dislocation) structure, which results in a short-time difference in the residual and maximal inductions of the loop.

The part played by defects in the nucleation process should particularly be noted. Defects are fundamentally necessary for originating the regions of back magnetization [11] enhancing the ferromagnet remagnetization. And though in view of (3.41), the noise level is proportional to the volume density of defects, it does not follow that the noise will be reduced to zero in the absence of defects. This is valid, first of all, for the Permalloy materials with low values of the constants of magnetostriction and magnetic anisotropy, in which a structure of optimum set of crystalline defects forms as a result of thermal processing (see Sect.3.5) and [29]). Since the defect structure is slowly rebuilt, slow changes in the nucleation processes can lead to short-time asymmetry of the hysteresis dynamic loop.

A critical field for nucleation is proportional to the density of the bound energy determined by the constants of magnetostriction and magnetic anisotropy [11, 29]. Therefore, by decreasing these constants, the noise emerging at the nucleation stage also decreases.

Thus, the decrease of the magnetostriction and magnetic anisotropy constants at the expense of variation in chemical composition and thermal processing is quite general, applicable to all stages of core remagnetization, and giving an effective means of suppressing the magnetic noise.

3.6 Other Relations and Dependencies

3.6.1 The Dependence on the Drive Field Frequency

With increasing the drive field frequency, the number of interactions per unit time of domain walls with defects grows. Therefore, the spectral density g of the noise EMF is proportional to the frequency $f_1 = \omega/(2\pi)$ of the drive field, $g \sim f_1$. However, in view of (3.20), the noise level $\overline{B}_n$ is dependent not only on g, but also on the transformation coefficient G. Thus we have

$$\overline{B}_n \sim f_1^{-1/2} . \qquad (3.43)$$

In practice, this proportionality usually holds true, the noise weakens with increasing frequency, but only if the core thickness satisfies (3.5) and until the surface effect comes into force. It is usual to choose the drive field frequency not higher than 20–30 kHz to save power in the excitation circuit and for other reasons.

The usual and the excess noises are sharply enhanced at harmonics of the core mechanical resonance. These undesirable effects can be avoided by

3.6. OTHER RELATIONS AND DEPENDENCIES

selecting the length of bar cores or the diameter of ring cores for a specific excitation frequency. The forbidden length or diameter are

$$l_{forbidden} = \frac{k}{4nf_1} c^{1/2}, \quad (3.44)$$

$$D_{forbidden} = \frac{\sqrt{c}}{2\pi n f_1}\sqrt{1+(1-k)^2},$$

where $n = 1, 2, 3, \ldots$ is the order of disturbing oscillations (with regard to the parity of the magnetostriction effect due to which the magneto-mechanical relation is realized), $k = 1, 2, 3, \ldots$ is the order of mechanical oscillations of cores at the time of their resonance, and c is the ratio of the elastic modulus w.r.t. the core material density.

3.6.2 Operational Temperature Dependence

The noise level is found to be approximately invariable in a wide temperature range, 100 to 350 K [21].

At low temperatures (<100 K), the noise is seen to be approximately 3 times greater than at room temperature. The noise level increase at low temperatures may be attributed to growing magnetic anisotropy.

At high temperatures, approaching the Curie point, the noise is decreasing. This was closely traced in the alloy Ni72Mo3Cu14Cr2Fe characterized by a low Curie point of $\Theta = 100 - 200°C$.

The dependencies of the noise level $\overline{B}_n(T)$, $B_S(T)$ and $\mu_{init}(T)$ [32] for the same alloy are compared in Fig. 3.11.

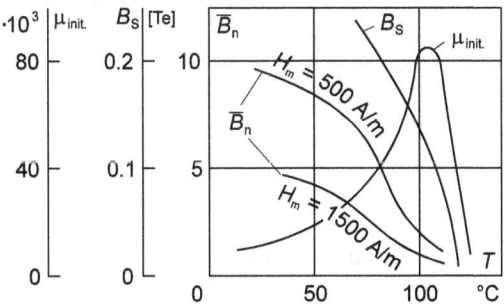

Figure 3.11: Temperature dependencies of $\overline{B}_n$, B_S and $\mu_{init.}$ for the alloy Ni72Mo3Cu14Cr2Fe.

We see that, approaching the Curie point, the saturation induction B_S decreases, the initial permeability $\mu_{init.}$ increases, and the peak in the curve of $\mu_{init.}$ corresponds to the minimum of $\overline{B}_n$.

Increasing $\mu_{init.}$ near the Curie point is consistent with decreasing the coefficient of rectangular form of the loop, B_R/B_S, which, as a whole, indicates a reduction of the magnetic anisotropy of the alloy.

Obviously, operating fluxgate cores in the vicinity of the Curie point is a method for lowering the noise. However, this can only be used in practice if materials with a Curie point $\Theta \leq 100°C$ are developed (see Sect. 3.7).

3.6.3 Spatial and Temporal Temperature Variations

The work [21] gives a brief review of the publications describing the relation of fluctuations in the parameters of the Barkhausen jumps to the temperature variations. The author suggests that this dependence reflects the "shaking down" of the energy states and the relation to altering the ferromagnet magnetic structure.

Another interpretation of this relationship is based on taking into account the rate of heat transfer in cores and, therefore, on the consideration of temperature gradient. The presence of temperature gradients can lead to both fluctuations in the parameters of the Barkhausen jumps and fluctuations in the thermal currents [37], which contribute to the excess noise. A direct relationship between the noise level and the temperature gradient along the fluxgate bar core was found by Saytzev and Gorobey. In some cases a nonuniformity of the internal drive field causes temperature gradients.

3.6.4 Volume, Shape and Processing Effects

It was determined [36] that the noise level $\overline{B}_n$ in magnetic amplifiers varies inversely with the square root of the core volume V,

$$\overline{B}_n \sim V^{-1/2}, \qquad (3.45)$$

or

$$\left.\overline{B}_n\right|_{l=const\neq 0} \sim s^{-1/2},$$

where s is the cross sectional area of the core, if the length l of the closed core is fixed.

Relations (3.45), when applied to fluxgates, as a rule, do not hold true (p. 351 in [36]). This is due to the fact that, in contrast to the transformation ratio of magnetic amplifier, the coefficient for fluxgate devices only slightly

3.6. OTHER RELATIONS AND DEPENDENCIES

increases with cross sectional core area s (at a given length l). It seems to be more important that the almost complete longitudinal noise correlation takes place in both halves of a closed core (see Sect.3.4). For this reason, the coefficient k in (3.22) and (3.23) for the fluxgate devices with closed cores is close to the coefficient of misbalance ε, that in view of (2.36), determines the extent of suppression of the coherent EMF induced in the sense coil.

The minimum value of the cross sectional area in fluxgate devices is found from the necessity of providing a high degree of uniformity of the internal drive field and, so, a high degree of longitudinal correlation of noise over the core length (see Sect. 3.4). The same should be said about choosing a core shape: the best form is one that maximizes the uniformity of the drive field.

Fluxgate cores are usually made from Permalloy tape of 0.02–0.1 mm thickness. Thinner tapes are not used because of the reduced magnetic qualities caused by plastic deformation (by increasing the volume density of defects) [29, 32]; thicker tapes are not used because of the appearance of the surface effect and increase of power expended for excitation.

Cores are manufactured in a press (applied only to isotropic materials) or by tape winding. In the latter case, the tape of a given width is cut with the help of a precise roller cutter. After cutting out a sample and threading a tape, it is deburred and then chemically or electrically polished. These procedures reduce the density of macroscopic defects which occur near the cut and affect the noise level.

Before winding, the tape must be degreased and covered by a heat-resistant insulator. These operations are often performed during the winding process. For insulation, electrolysis is applied and the tape is coated with different suspensions.

During electrolysis, an insulating substance being in a suspension state is applied to the tape surface electrostatically. Use is made of different suspensions: silicon dioxide in acetone, magnesium oxide in carbon tetrachloride, ammonium oxide in methyl alcohol and others. The tape in the electrolysis is transported at 1 to 3 m/min. The covering thickness is regulated by changing the speed, the suspension concentration, and the applied voltage.

The permalloy cores wound on bobbins are annealed together. The heat-resistant materials for the supports are made from nonmagnetic alloys, for example the alloy NiMo+X[4], "chrovangal", and others as well as ceramic materials, glass ceramic, and quartz glass. The supports must have a low mass (for a low thermal inertia), provide a reliable protection of the Permalloy

---

[4]NiMo+X is a non-magnetic alloy with a high specific resistance $\rho = 1.5 - 1.6 \; \Omega mm^2 m^{-1}$, a high melting temperature (1350°C) and a linear expansion coefficient close to that of permalloy ($\alpha = 12100^{-6}$ K^{-1}).

wound core from external mechanical stresses, possess a thermal coefficient of linear expansion close to that of the Permalloy, and minimize the drive coil to core separation.

The last turn of the Permalloy tape is connected to the previous turn by using spot or laser welding. It should be remembered that the welding point is a macroscopic defect which reduces the uniformity of the internal field in the core. If welding connects only two turns of the tape, then other turns "shunt" this defect (see Sect.3.4). Therefore care must be taken lest welding punches through additional turns of the core.

Preliminary mechanical processing of the cut tape, its covering by a heat-resistant insulator, further winding and fixing on the support are fairly important and responsible procedures determining the level of the magnetic noise. Annealing is the last operation affecting the noise level. Note that the best results are achieved by annealing cores in hydrogen.

3.7 Prospects for Suppressing Noise

The lowest noise level is achieved in fluxgates with cores made mainly from molybdenum permalloy. Typical values include [54, 64, 69]:

- the average value of noise, $\overline{B}_n = 7 - 12$ pT in the frequency range 0.01 to 1 Hz;

- the spectral density of the noise is, $b_n = 3 - 5$ pT$/\sqrt{Hz}$ beginning from 1 Hz up (see Sect. 3.3).

The above-mentioned noise values are achieved on using the native Permalloy materials Ni83V5Fe, Ni82Mo6Fe and Ni81Mo at the maximum excitation, $h_{max} = 2-3$ kA/m. In some cases, the upper bound of the noise is downward biased, and the uniform magnetic noise of density b_n is observed beginning at fractions of 1 Hz [73]. Yu. N. Bobkov and E. N. Langvagen informed the author of similar measurements.

The work [66] briefly reports on a fluxgate instrument designed for medicinal purposes having a sensitivity threshold of the order of 0.3 nT at 3 Hz. It was noted that the new molybdenum Permalloy was used in these instruments, but particular features of the instrument design were not reported. Therefore, it remains unclear whether this threshold was reached by employing the new low-noise permalloy, or the fluxgate devices were additionally instrumented with inert (for example, ferrite) concentrators enhancing the field near the fluxgate units .

Below, we consider possible ways to lower the noise and thereby the sensitivity threshold of fluxgate devices only by using new materials.

3.7.1 Magnetic Material Parameter Selection

It follows from the above that, all other things being equal (the same extent of core saturation, uniformity of the drive field, quality of manufacture, etc.), the level of magnetic noise is dependent on the magnetically soft materials used. Following the [8], the material parameters are divided into structurally insensitive and structurally sensitive ones. The nonuniformity in a crystal lattice are due to impurities, vacancies, micropores, dislocations, residual stresses and so on. The structurally insensitive parameters involve the magnetic anisotropy factor K, saturation magnetostriction λ_S, the saturation induction B_S (magnetization J_s) and the Curie point Θ; the structurally sensitive parameters are the permeability μ, the coercive force H_C, and the hysteresis loss W_H.

The magnetic noise level should seemingly be related, first of all, to the structurally sensitive material parameters. However, the character of such a relation turns out to be fairly contradictory. For example, the minimum noise level is not achieved with materials characterized by the greatest value of μ_{max} (alloy Ni79Mo3.9Cu-0.2Fe, Supermalloy and others). It does not always happen that the minimum noise is associated with the smallest coercive force H_C in the materials (table3.1).

At the same time, as was shown in Sect. 3.5, there is a clear relationship between the noise level and structurally insensitive parameters such as the magnetic anisotropy factor K and the saturation magnetostriction λ_S. It is desirable for these parameters to be minimized together (see Fig. 3.10). Furthermore, by doing this minimization, we also obtain useful changes in such structurally sensitive parameters as the initial permeability, $\mu_{init.}$, the factor of rectangular form of the loop, B_R/B_S, and the coercive force H_C.

Detailed criteria for low noise materials are:

$$\left.\begin{array}{ll} K \to 0, & \lambda_S \to \quad\quad B_R/B_S \to 0.5, \\ \mu_{init.} \to \max, & H_C \to W_H \to \quad D \to \max, \end{array}\right\} \quad (3.46)$$

where D is the material hardness. One sees that the criteria involve both the structurally sensitive and insensitive parameters.

Criterion (3.43) is satisfied by the above-mentioned crystalline iron-nickel alloys Ni83V5Fe, Ni82Mo6Fe and Ni81Mo, as well as by magnetically soft amorphous alloys (see below).

3.7.2 Curie Point Dependence

In Section 3.6, referring to the work by Afanasenko and Berkman, we noted that the operation of fluxgate cores, i.e. magnetic amplifiers, near the Curie point is a means for obtaining a minimal noise level. It is to be pointed out that, even earlier, Kolachevskiy [21] also observed a lowering of the noise level at temperatures closer to the Curie point Θ.

Shiray [72] performed another experiment. He did not vary the temperature to reach the Curie point, but conversely, the Curie point was varied by adding molybdenum and chromium to amorphous alloys so as to reach room temperature. The alloy $Fe_{78}Si_{12}B_{10}$ was utilized in the experiment. On addition of Mo and Cr to the alloys in the ratio $(Fe_{78}Si_{12}B_{10})_{100-x-y}Cr_xMo_y$, the Curie point Θ was lowered from 400 to 70°C, an exponential decrease of the noise level being also observed. The results presented in [72] demonstrate not only a reduction of K and λ_S, but also minimization of the value of B_s. The noise level is proportional to the hysteresis level, $W_H = H_C B_S$, and so becomes weaker not only with decreasing the coercive force H_C, but also with decreasing the saturation induction B_S.

Note, also, that decreasing B_S gives rise to a higher stability of the fluxgate zero position, since the absolute level of ghost even harmonics voltage is lowered (see Sect. 3.2.1).

With regard to (3.46), the relationship between material parameters resulting in a low noise level on lowering the Curie point can be represented in the form

$$\left.\begin{array}{l|l|l}
T \to 0 & K \to 0, & H_C \to \\
\text{or} & \lambda_S \to 0, & \mu_{init.} \to \max, \\
\Theta \to & B_S \to & B_R/B_S \to 0.5, \\
\text{(up to 50\ldots 100°C)}, & \text{(up to 0.1\ldots 0.2 T)}, & W_H \to
\end{array}\right\} \quad (3.47)$$

Of course, we are dealing not with attaining a temperature of 300 − −400 in fluxgates using the low-noise alloys Ni83V5Fe, Ni82Mo6Fe and Ni81Mo (from the practical standpoint, it is absolutely unacceptable), but with the need to develop materials having the same parameters and additionally a low Curie point Θ that, for example, lies within the temperature range 50 − −100°C. This allows for a low noise level to be reached in fluxgates.

As was noted in [8], the low Curie point in the molybdenum Permalloys can also be obtained by increasing the molybdenum content. Other melt compositions are known to lower the Curie point. An example is given by the alloy Ni72Mo3Cu14Cr2Fe [32]. Considerable developments of amorphous alloys offer new possibilities. In addition to satisfying criterion (3.47), amorphous alloys possess other positive properties improving the technical parameters of fluxgate devices and lowering the cost of these instruments. Therefore, primary attention should be paid to developing new low-noise amorphous alloys.

3.7.3 Amorphous Materials

The lowest noise level obtained up to now [72] argues for developing and studying amorphous alloys. It was shown [72] that the smallest noise level is characteristic of amorphous alloys with a cobalt content of about 67% (in this case the condition $\lambda_S \to 0$ is reached). Shiray also found that, at such cobalt content, the noise is minimal provided that the alloy contains 3% iron. On addition of chromium to the alloy, further noise reduction is observed (accompanied by lowering the Curie point Θ) (see Fig. 3.12).

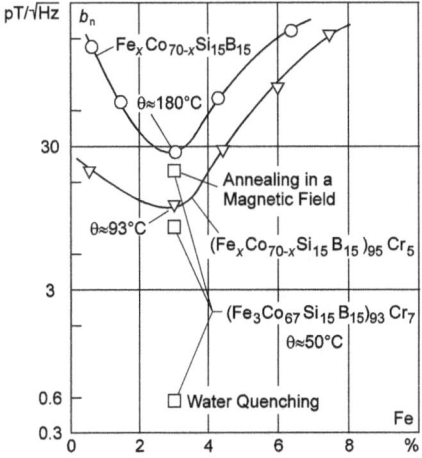

Figure 3.12: Dependence of the noise level b_n on the composition of the amorphous alloy and on the regime of sample cooling after thermal processing.

By using the alloy $(Fe_3Co_{67}Si_{15}B_{15})_{93}Cr_7$ and after water quenching, Shiray determined the noise level to be $b_n = 0.6$ $pT/\sqrt{Hz}$ (in the frequency range 0.1 to 16 Hz). It follows from this data (see Fig. 3.12) that due to water quenching, the noise was reduced more than 20 times! Furthermore, it was noted that the internal structure of samples acquired a rod structure with complex domains different from simple domains (with the 180°$walls$).

Shiray also showed that the same material being annealed in a magnetic field produces a high noise (see Fig. 3.12), which seems to be related to enhancing the induced anisotropy and to satisfy criterion (3.47). It follows from the preceding that the typical values of the noise level given at the beginning of Sect. 3.7 are not limiting values, and there are real possibilities for further reduction of the noise level in fluxgate units, primarily by development of new materials and thermal processing methods.

Chapter 4

Functional Design and Error Calculation

4.1 Functional Design

4.1.1 Magnetometers for Constant Fields

The most widely used design for a magnetometer utilizes "detection of the second harmonic". This provides the advantages of the second mode of fluxgate operation (see Chap. 1). The standard design of a magnetometer is depicted in Fig. 4.1.

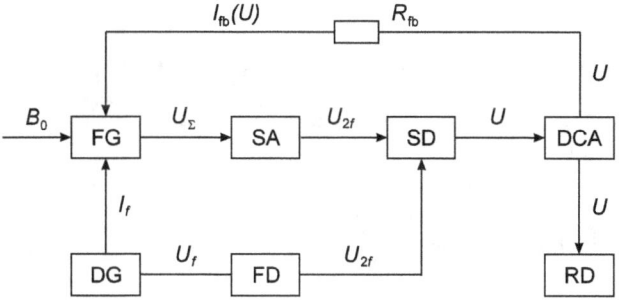

Figure 4.1: The design of a magnetometer for DC field measurement: FG – fluxgate; SA – selective amplifier SD – synchronous detector; DCA – direct current amplifier; DG – drive generator; FD – frequency doubler; RD – registration device; R_{fb} – feedback resistor.

In this design the standard method of modulation, amplification and demodulation of the weak constant or slowly changing (infra-low-frequency) signals occurs. A fluxgate producing a modulated signal of carrier frequency $2f$ serves as a modulator. This signal is then amplified. As a rule, with a view to improving the ratio of signal $2f$ from the odd harmonic interference (misbalance voltage) selective amplification is used. A synchronous detector serves as a demodulator. The detector and modulator are synchronized with a signal from a common oscillator. The demodulated signal is further amplified by a DC amplifier. This signal attains a level where reliable measurement of the output voltage or current is possible.

The synchronous detector is an important element in the design, in contrast to ordinary "linear" detectors it is not only a demodulator but it is also phase and frequency sensitive. If the integration time of a demodulated signal is increased it is possible to achieve an extremely narrow channel bandwidth for the magnetometer. Synchronous detection is widely used for the suppression of internal interference (odd voltage harmonics caused by fluxgate mismatch) and external interferences as well, for example, the parasitic fields of industrial frequency 50 Hz, and its divisible frequencies. The integrator can be realized by RC-circuitry in the synchronous detector and a direct-current amplifier.

As a rule, with a view to stabilize the whole transformation ratio (amplification) a magnetometer channel is covered by negative feedback on the field, which is achieved by feeding of some part of the output current of a channel through the feedback resistor to the compensation coils (see Sect. 4.3). Sometimes, channel elements themselves (for example, amplifiers) have this negative feedback.

4.1.2 Magnetometers for Alternating Fields

Fluxgates are characterized by practically the same transformation ratio when applied to both DC and AC magnetic fields [3, 28].

For this reason fluxgate magnetometers for the simultaneous measurement of DC and AC fields can be created. It is necessary to choose the channel bandwidth of the magnetometer to allow amplification and transformation of the whole frequency spectrum of a modulated signal. The design of a magnetometer remains as it is shown in Fig. 4.1.

But in the case where only alternating fields are to be measured it is better to use the design of a magnetometer shown in Fig. 4.2. Essentially, this diagram contains the same elements as in Fig. 4.1. The only difference is that now part of the channel has frequency-dependent negative feedback on the field and instead of the DC amplifier a selective amplifier of a low frequency (signal frequency) and a linear detector are used.

The reason for introducing frequency-dependent negative feedback is to

4.1. FUNCTIONAL DESIGN

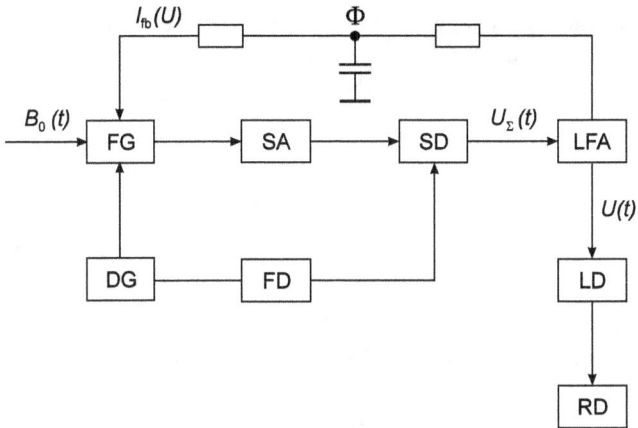

Figure 4.2: The schematic for an AC magnetometer: LFA – low frequency amplifier; Φ – frequency dependent filter; LD – linear detector (the remainder is as in Fig.4.1.).

compensate a constant geomagnetic field within fluxgate volume, and to provide, by having created favorable measurement conditions, amplification and transformation of the weak signals of an alternating field.

A selective low frequency amplifier with a given bandwidth is used to optimally filter the signal. For narrow bandwidth signals a low-frequency amplifier can be superimposed not with a linear but with a synchronous or asynchronous detector [3, 4].

There are no difficulties in creating a fluxgate magnetometer with an upper frequency range of several kilohertz and higher (it is limited only by the frequency of the excitation field). But at frequencies higher than 100 Hz passive induction magnetometers begin to compete with fluxgate magnetometers [28]. The most useful operating frequency for fluxgate magnetometers is of the order of up to some tens of Hz.

In comparison with passive induction magnetometers fluxgate magnetometers have the following advantages:

1. they directly measure the value B_{0i} (or H_{0i})[1], where $i = x, y, z$, and not the derivative dB_{0i}/t ;

2. the amplification of a signal occurs at a higher frequency (owing to modulation), this results in a high signal-to-noise ratio at the input to

---

[1] Here and further on vector $\mathbf{B}_0$ and its projections B_{0i} are used (see Sect. 2.1).

the electronics of the magnetometer (amplifier noise increases with a reduction in the frequency of the input signal), minimal phase distortion of the signal and more stable operation of the device as a whole;

3. the same coefficient of fluxgate transformation makes it possible to use the same channel for simultaneous or non-simultaneous measurement of alternating fields of different frequencies.

4.1.3 Magnetic Gradiometers

Using two fluxgates it is possible to create a device for measuring the difference of a magnetic field at two points in space: $\Delta B_{0i} = B_{0i}'' - B_{0i}'$. If the distance between the magnetic centers of fluxgates (base) is comparatively large then the device is called a differential magnetometer; if the base is small, then the device is called a magnetic gradiometer. For a gradiometer it is possible to take $\Delta B_{0i}/\Delta l \approx \partial B_{0i}/\mathrm{d}l_j$, where l is length, i and j are indices of directions along the appropriate coordinate axes.

But independently of the nomenclature, the design of the device is the same (Fig. 4.3). One can see that the device contains two magnetometer channels, one of which (the upper one in the figure) is used for the compensation of the homogeneous part of the field in the region of the two fluxgates.

The designs considered above are used in devices for different purposes with some additions (see chapter 8). They are all based on the magnetometer design depicted in Fig. 4.1. The metrological characteristics of any fluxgate device depend much on how well this design is made, calculated and realized. For this reason the analysis of device errors will be started with this design exactly.

4.2 Total Error

The purpose of error analysis is the estimation of the resultant or total error of a device caused by the errors in its component parts. First of all an estimation of the basic and additional errors of a device is made.

For the estimation of a basic error (device errors in normal conditions) the method of structural analysis is used. This allows one to follow the errors of individual units and blocks to the input or output of a device. In the practice of magnetic measuring it is accepted to follow these errors to the input of a device. And the total error can be directly expressed as absolute error ΔB_{0i} or as given (relative to the upper level of measurement range B_{lim}) $\delta = B_{0i}/B_{\mathrm{lim}}$.

4.2. TOTAL ERROR

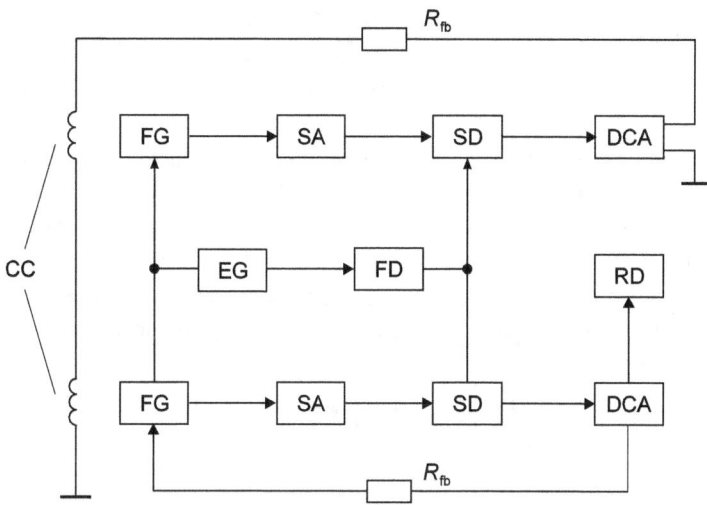

Figure 4.3: Diagram of a gradiometer (a differential magnetometer): CC – compensatory coils of a homogeneous field which are put on fluxgates (the remainder is as in Fig.4.1).

Additional errors, occurring with the deviation of measurement conditions from the norm, are normalized as the ratio of the absolute error to the influenced quantity causing this error, or as the ratio of a relative error to the same influenced quantity. According to new standards it is recommended to express an additional error as a percentage of the basic error, but this is not always convenient.

The frequently used term "class of device accuracy" characterizes the relative error. Consequently, the accuracy class characterizes the maximal assumed error of a device, but it is not the direct characteristic of its real error or accuracy.

The exposure of its multiplicative and additive components is of great practical importance for the total error minimization of a device when it is designed, produced and tuned.

Errors are defined as multiplicative if they depend on the value and direction of measured field vector $\mathbf{B}_0$. If the components of an error do not depend on vector $\mathbf{B}_0$ they are defined as additive. The reason for seperating the error into these parts is that there are general methods for the minimization of these components, and consequently, of the device total error [4, 40].

The transformation equation for a single component magnetometer (Fig. 4.1) is as follows:

$$U_i = S_i B_0 \cos \alpha_i + U_{0i}, \tag{4.1}$$

where U is the output voltage; S is the transformation coefficient; B_0 is the value (scalar) of a measured field vector; $\alpha = \mathbf{B}_0, \mathbf{i}_M^0$ is the angle between vector $\mathbf{B}_0$ and the device magnetic axis (see Sect. 2.7) U_0 is the voltage, characterizing the zero offset and channel noise and $i = x, y, z$.

If $\mathbf{B}_0 = const \neq 0$ the values of quantities S, α and U_0 do not remain constant. Their changes define device errors. The total error at the output is

$$\Delta U_i = \frac{\partial f}{\partial S_i} \Delta S_i + \frac{\partial f}{\partial \alpha_i} \Delta \alpha_i + \frac{\partial f}{\partial U_{0i}} \Delta U_{0i}. \tag{4.2}$$

Expressing the increments of angle $\Delta \alpha_i$ through angles $\Psi_{ij}(i \neq j)$ – see Fig. 4.4 – taking into consideration small values Ψ_{ij} and U_0, calculating particular derivatives and passing on to total errors at the output as applied to a three-component magnetometer, we find [4, 40]:

$$\left. \begin{array}{rcl} \Delta B_{0x} & = & \dfrac{\Delta S_x}{S_x} B_{0x} + \Psi_{xy} B_{0y} + \Psi_{xz} B_{0z} + \dfrac{U_{0x}}{S_x}, \\[6pt] \Delta B_{0y} & = & \dfrac{\Delta S_y}{S_y} B_{0y} + \Psi_{yz} B_{0x} + \Psi_{yz} B_{0z} + \dfrac{U_{0y}}{S_y}, \\[6pt] \Delta B_{0z} & = & \dfrac{\Delta S_z}{S_z} B_{0z} + \Psi_{zx} B_{0x} + \Psi_{zy} B_{0y} + \dfrac{U_{0z}}{S_z}. \end{array} \right\} \tag{4.3}$$

The first three components on the right side of each line are the multiplicative components of the error, which are proportional to the corresponding component value of the measured field vector $\mathbf{B}_0$. The last component in each line is the additive component of an error which does not depend on the field of vector $\mathbf{B}_0$.

The summation of error components is made according to their nature. If ΔS, Ψ and U_0 are deviations from nominal values which have a systematic nature, i.e. are observed steadily for a long period, summation is made according to formulae (4.3). If these deflections are statistically (mean, mean square) distributed, for example, according to a normal law, then summation

4.2. TOTAL ERROR

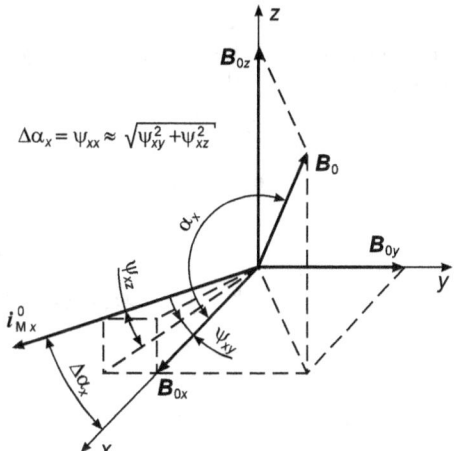

Figure 4.4: Vector diagram explaining the introduction of angles Ψ_{ij} (for simplicity only the x-channel is shown).

is carried out on formulae

$$\left.\begin{aligned}
\Delta B_{0x} &= \sqrt{\left(\frac{\Delta S_x}{S_x}\right)^2 B_{0x}^2 + \Psi_{xy}^2 B_{0y}^2 + \Psi_{xz}^2 B_{0z}^2 + \left(\frac{U_{0x}}{S_x}\right)^2}, \\
\Delta B_{0y} &= \sqrt{\left(\frac{\Delta S_y}{S_y}\right)^2 B_{0y}^2 + \Psi_{yx}^2 B_{0x}^2 + \Psi_{yz}^2 B_{0z}^2 + \left(\frac{U_{0y}}{S_y}\right)^2}, \\
\Delta B_{0z} &= \sqrt{\left(\frac{\Delta S_z}{S_z}\right)^2 B_{0z}^2 + \Psi_{zx}^2 B_{0x}^2 + \Psi_{zy}^2 B_{0y}^2 + \left(\frac{U_{0z}}{S_z}\right)^2}.
\end{aligned}\right\} \quad (4.4)$$

In Fig. 4.5 the changing nature of the multiplicative and additive components of an error for the linear graduated characteristic of a magnetometer is shown.

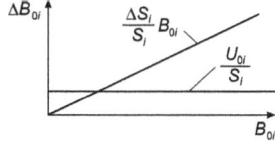

Figure 4.5: The changes of a multiplicative and additive error in the range of measured values (for the linear graduated characteristic of a magnetometer).

One can see that when relatively strong magnetic fields are measured multiplicative errors are dominant and the additive errors can be neglected. On the contrary, when weak magnetic fields are measured the additive error is dominant and multiplicative errors can be neglected.

These circumstances must be taken into account when devices are designed, especially as the methods of the minimization of these or other errors are quite different.

4.3 Methods for the Minimization of Multiplicative Errors

4.3.1 Errors of Transformation Ratio

According to formulae (4.3), (4.4) and Fig. 4.5 the multiplicative error which was caused by the deviation

$$\Delta S_i = S_i - S_i^*$$

of transformation ratio δS_i from its nominal value S_i^* is proportional to the measuring field component B_{0i}.

It is possible to reduce this error by nulling the field component B_{0i} at the fluxgate. Two block diagrams for field nulling are shown in Fig. 4.6. The nulling is achieved by supplying current into a compensation coil which is mounted on the fluxgate. In the first diagram (Fig. 4.6a) compensation current is formed by a feedback loop (static regulation conditions). In the second diagram (Fig. 4.6b) current is supplied from an autonomous source setting the amount of compensation according to a zero-indicator connected with a channel (astatic regulation conditions).

In both diagrams it is essential to distinguish a chain of *direct* transformation from a chain of *inverse* transformation. According to the accepted signs a chain of *direct* transformation will be called k-chain, a chain of *inverse* transformation β-chain.

Error minimization on the first diagram is achieved if

$$S = \left.\frac{k}{1 + k\beta}\right|_{k\beta \gg 1} \approx \frac{1}{\beta} = \frac{R}{C_K}, \qquad (4.5)$$

where $k = \sum_{m=1}^{n} k_m$ - direct transformation coefficient; k_m - coefficient of the transformation of some k-chain elements ($k_1 = G$ - coefficient of fluxgate

4.3. METHODS FOR THE MINIMIZATION OF MULTIPLICATIVE ERRORS

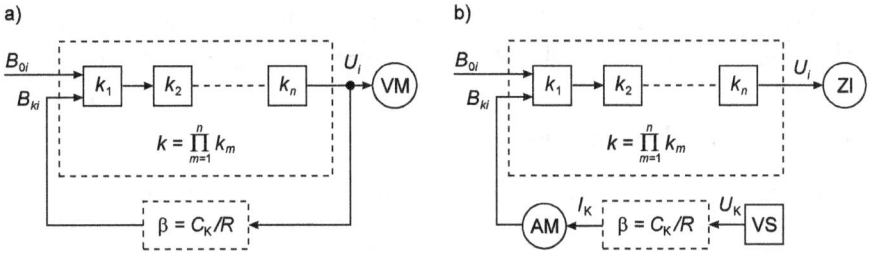

Figure 4.6: Magnetometers with feedback (a) and compensation (b): k_i – coefficient of transformation of the k-chain elements; C_k and R – coefficients of transformation of the β-chain elements; VM – voltmeter; AM – amperemeter; ZI – zero-indicator; VS – voltage source.

transformation); $\beta = C_K/R$ - coefficient of inverse transformation; C_K - compensation coil constant; R - total resistance of a feedback chain.

From expression (4.5) we have

$$\frac{\Delta S}{S} = \frac{\Delta R}{R} - \frac{\Delta C_K}{C_K}. \qquad (4.6)$$

Such a relative error of a channel with feedback exists if output voltage is subject to direct measurement.

If, instead of voltage, feedback current is subject to direct measurement the relative error of a transformation coefficient will be

$$\frac{\Delta S}{S} = -\frac{\Delta C_K}{C_K}, \qquad (4.7)$$

i.e. it is defined by the error of the compensation coil constant.

The minimization of the multiplicative error in the second diagram is achieved if $B_{0i} = -B_{ki}$, where B_{ki} is a compensatory field.

In this case k-chain serves as a zero-indicated chain, either voltage or current in β-chain is subject to direct measurement and, consequently, errors are carried in by the elements of this chain.

If, at the moment of field compensation, voltage is registered on the resistor R, the relative error of the transformation coefficient of the magnetometer is defined by expression (4.6). If the compensation current is measured, then the error is defined by expression (4.7).

It follows from this that there is no fundamental difference in these methods of the $\Delta S/S$ error minimization; either of these methods in fact involves the replacement of the less stable k-chain which contains the active elements

– a fluxgate, electronic amplifier and transformer – with a more stable β-chain which contains the passive elements - compensation coil and a resistor.

4.3.2 Angular Errors

Multiplicative errors caused by the deviation of a fluxgate magnetic axis from the given direction i.e. connected with the presence of the Ψ_{ij} angles, can be minimized by the methods described in Sect. ??. To these methods the following ones can be referred:

- the choice of cores with the smallest diametrical permeability;

- firm attachment of all the elements of a fluxgate and minimization of the angle of misalignment between the core longitudinal axis and the axis of the sense coil;

- stabilization of the amplitude (or maximum value) of the excitation field.

From formulae (4.3) and (4.4), multiplicative errors are proportional to the values of the diametrical components of measured field $\mathbf{B}_0$. Thus, in this particular case, the best method for minimization of errors will be the compensation of these field components at the fluxgate.

As mentioned above, the condition for error minimization connected with the instability of the transformation ratio of a magnetometer is the following equality

$$B_{0i} + B_{ki} \approx 0 . \qquad (4.8)$$

Here B_{0i} and B_{ki} are projections of vectors $\mathbf{B}_0$ and $\mathbf{B}_k$ ($\mathbf{B}_k$ is the vector created by the compensation coil) on the fluxgate magnetic axis (Fig. 4.7). We shall call this compensation of the field *scalar*. One can see that, with scalar compensation, the diametrical components of field $\mathbf{B}_0$ are not compensated. What is more, if the axis of the compensation coil is not aligned with the magnetic axis of the fluxgate (as shown in the figure), the additional diametrical components of field B_{kj} appear, they influence adjacent fluxgates in three-component magnetometers, and lead to one more type of a multiplicative error.

It is evident that the minimization of all multiplicative errors of a magnetometer will be realized if the following condition is fulfilled

$$\mathbf{B}_0 + \mathbf{B}_K \approx 0 . \qquad (4.9)$$

4.3. METHODS FOR THE MINIMIZATION OF MULTIPLICATIVE ERRORS

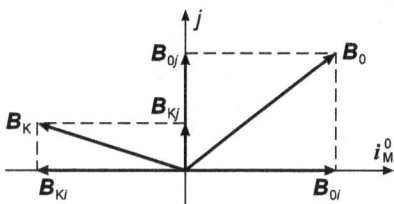

Figure 4.7: Vector diagram explaining scalar compensation of the field at the fluxgate.

Such compensation of a field will be called vector compensation. With vector compensation even errors caused by the uncertain spatial position of the magnetic axes of a fluxgate are minimized.

This was first demonstrated by Afanassiev and Bushuev and later by Primdahl and Anker-Jensen [65]. The magnetometer design by Afanassiev and Bushuev is shown in Fig. 4.8.

The fluxgate block, including three single-component fluxgates or a single three-component fluxgate, is set inside a cube which serves as a framework for the three-component system of compensation coils. The functional connection of all the elements is evident from the figure. The signals of the second harmonic U_{2i} produced by a fluxgate block are fed to the inputs of the appropriate electronic channels. Autocompensation currents I_{ki} produced by every channel are fed to the appropriate compensation coils. As a result, with rather strong feedback, condition (4.9) is fulfilled, i.e. the vector compensation of a field at the fluxgate block is provided.

If vector compensation is fulfilled the second and third components $\Psi_{ij}B_{0j}$ in each line of (4.3) should be changed into $\Psi_{ij}\delta B_{0j}$, where δB_{0j} is the field of undercompensation in the direction appropriate to component B_{0j} of vector $\mathbf{B}_0$,

$$\delta B_{0j} = B_{0j} - B_{Kj} . \tag{4.11}$$

If we do not take into account the second component of ratio (4.1), with regard for (4.5) we can get instead of (4.11)

$$\delta B_{0j} = (1 - \beta S_j)B_{0j} = \frac{1}{1 + k_j\beta_j}B_{0j} . \tag{4.12}$$

For this reason

$$\Psi_{ij}\delta B_{0j} = \frac{\Psi_{ij}}{1 + k_j\beta_j}B_{0j} . \tag{4.13}$$

From this ratio one can see that errors connected with the angles Ψ_{ij} are reduced $(1 + k_j\beta_j)$ times. If feedback is very strong (for example, $k\beta >$

CHAPTER 4. FUNCTIONAL DESIGN AND ERROR CALCULATION

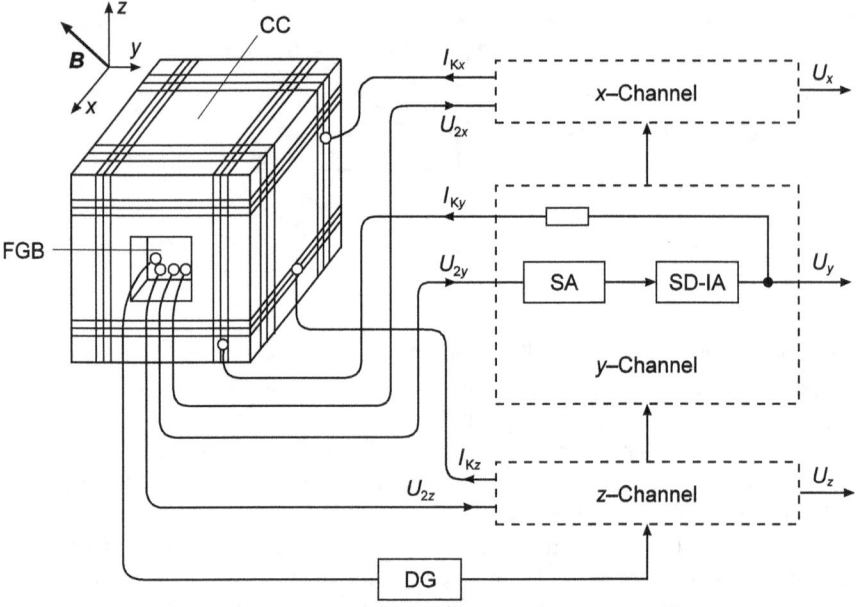

Figure 4.8: Diagram of a three-component magnetometer with vector compensation of the field: FGB – block of fluxgates; CC – three-component system of compensatory coils on a cubic frame-work; SA – selective amplifier; SD-AI – synchronous detector with an integrating amplifier; DG – drive generator.

100) the axes of the compensation coils become the magnetic axes of the magnetometer. The changes of the spatial position of the fluxgate magnetic axes, i.e. angles Ψ_{ij}, are not important, whatever the cause, this is the advantage of the vector compensation over scalar compensation of the field.

For a three-component magnetometer with vector compensation of the field (Fig. 4.8) with the same strong feedback ($k\beta > 100$) in each channel

4.3. METHODS FOR THE MINIMIZATION OF MULTIPLICATIVE ERRORS

instead of (4.4), for example, we can write:

$$\left.\begin{aligned}\Delta B_{0x} &= \sqrt{\left(\frac{\Delta C_K}{C_K}\right)_x^2 B_{0x}^2 + \frac{1}{k^2\beta^2}\left(\Psi_{xy}^2 B_{0y}^2 + \Psi_{xz}^2 B_{0z}^2\right) + \left(\frac{U_{0x}}{S_x}\right)^2},\\ \Delta B_{0y} &= \sqrt{\left(\frac{\Delta C_K}{C_K}\right)_y^2 B_{0y}^2 + \frac{1}{k^2\beta^2}\left(\Psi_{yx}^2 B_{0x}^2 + \Psi_{yz}^2 B_{0z}^2\right) + \left(\frac{U_{0y}}{S_y}\right)^2},\\ \Delta B_{0z} &= \sqrt{\left(\frac{\Delta C_K}{C_K}\right)_z^2 B_{0z}^2 + \frac{1}{k^2\beta^2}\left(\Psi_{zx}^2 B_{0x}^2 + \Psi_{zy}^2 B_{0y}^2\right) + \left(\frac{U_{0z}}{S_z}\right)^2}.\end{aligned}\right\}$$
(4.14)

In conclusion, we should note that in magnetometers with vector compensation it is important to use one three-component magnetometer with a common magnetic center instead of three single-component magnetometers (see Sect. 2.7.1). This makes it possible to considerably reduce the overall dimensions of the compensation coils. Thus, the system of coils used by Afanassiev and Bushuev which was placed on the sides of a cube with a three-component fluxgate inside, had overall dimensions $40 \times 40 \times 40$ mm with the heterogeneity of the compensation field at the fluxgate coil of approximately 1–2%.

4.3.3 Errors Caused by Feedback Coils

It follows from this that the minimization of multiplicative errors in fluxgate magnetometers actually results from the fact that direct transformation is replaced by inverse transformation connected with the production of a compensatory field $\mathbf{B}_K$ at the fluxgate. This means that, at least for relatively strong magnetic fields, when additive errors can be neglected (Fig. 4.5), the errors of fluxgate magnetometers can be approximated to the error of magnetic induction coils current which has order 10^{-4}–10^{-5}, or less [40].

Of course techniques of stabilizing coil parameters originally developed for magnetic induction standards can be applied to fluxgate compensation coils.

It was shown above that the $\Delta S/S$ relative error of a magnetometer with field compensation becomes $\Delta C_K/C_K$, i.e. to the stability of the compensation coils. The temperature of the surroundings has the largest influence on the constant of a coil. This is why when compensations to be used, first of all thermostability of the coils should be taken into account.

When coils to measure magnetic induction are to be used the same task is solved in the following way. Hollow quartz cylinders outside of which special grooves are carved, are used as the frames for the coils. Then, in these grooves, an uninsulated copper wire is wound at high temperature under tension. As a result of this production method the error is reduced to a level close to the temperature coefficient of linear expansion of quartz ($\leq 5 \cdot 10^{-7}$ K^{-1}) [40, § 13-2].

A similar method can be used for producing the compensation coils of single-component fluxgates. Glass, electrotechnical porcelain or ceramics with smaller temperature expansion coefficients than copper wire can be used as thermostable materials.

For an axial field in the center of a single layer solenoid compensation coil [40]

$$B_K = \mu_0 I w (L^2 + D^2)^{-1/2}, \tag{4.15}$$

where μ_0 is the magnetic constant, I is the compensation current; w is the quantity of coils of solenoid; L and D are the length and the diameter of solenoid. If we make the L and D different (4.15), introduce designation $L/D = \lambda$ and come to relative increments we derive 0

$$\frac{\Delta B}{B_K} = \frac{\Delta C_K}{C_K} = -\left(\frac{\lambda^2}{\lambda^2+1}\frac{\Delta L}{L} + \frac{1}{\lambda^2+1}\frac{\Delta D}{D}\right). \tag{4.17}$$

From this one can see that changes of L and D do not have the same influence on $\Delta C_K/C_K$. If $\lambda \gg 1$

$$\frac{\Delta C_K}{C_K} = -\left(\frac{\Delta L}{L} + \frac{1}{\lambda^2}\frac{\Delta D}{D}\right), \tag{4.18}$$

i.e. relative deflection $\Delta C_K/C_K$ depends mainly on the relative change of solenoid length.

Solenoid length L can be stabilized by putting the winding into a spiral groove, carved on the surface of a cylindrical frame which is made from the thermostable materials mentioned above. As a result, it is possible to achieve relative error $\Delta C_K/C_K$ which is considerably less than the temperature coefficient of the linear expansion of a copper wire.

Let us make the following calculation.

First, let us assume that the solenoid coils are wound very close to one another and placed on the smooth cylindrical surface. Then, when temperature varies, the length of solenoid will change independently of the length of the cylinder. The temperature coefficient of linear expansion of copper wire is equal to $1.7 \cdot 10^{-5}$ K^{-1}. If $\lambda > 5$ then the second component (4.18) can be neglected. In this case we will have

$$\frac{\Delta C_K}{C_K}(T) \approx -\frac{\Delta L}{L}(T) \approx 1.7 \cdot 10^{-5} \text{ K}^{-1}.$$

4.4. MEANS FOR ADDITIVE ERROR MINIMIZATION

Now let us put the wire into a helical groove, for example, carved on the surface of a ceramic cylinder which has a temperature coefficient of linear expansion $1 \cdot 10^{-6}$ K^{-1}. In this case the length of solenoid will be stabilized, i.e. will have a thermal expansion coefficient equal to that of ceramics; the diameter of the solenoid will have a thermal expansion coefficient equal to that of copper. This is why, in accordance with (4.18), for $\lambda = 5$ we derive

$$\frac{\Delta C_K}{C_K}(T) \approx -\left(1 \cdot 10^{-6} + \frac{1.7}{25} \cdot 10^{-5}\right) \approx -1.7 \cdot 10^{-6} \text{ K}^{-1},$$

i.e. the temperature dependent error of the compensation coil is one order less. It is evident that if $\lambda > 5$ this error will approach 1 ppm, i.e. the coefficient of ceramics.

A small temperature dependence of the compensation coils with any λ can be attained by heating wire coils on the ceramic frame or by spraying the copper onto the frame just by the wire tension.

For stabilization of the spatial position of magnetometer axes, the frame for the coils must be monolithic and have an isotropic shape (a cube or a sphere).

In conclusion, we will point out another method for the minimization of errors caused by the temperature changes. This is the use of the block of fluxgate sensors with compression coils inside a thermostabilized chamber [49, 61].

4.4 Means for Additive Error Minimization

Additive errors (which are independent of the measured field) are caused by both internal and external noise, influencing different elements of the magnetometer.

Suppose, that some false signal U_p^f influences the input of the p-channel element of the magnetometer (see Fig. 4.6). Then the false signal in the feedback channel output is

$$U_{0i} = U_p^f \frac{\prod\limits_{m=p}^{n} k_m}{1 + k\beta}.$$

The additive error, considering (4.3) and (4.5) will be:

$$(\Delta B_0)_{ap} = \frac{U_{0i}}{S} = U_p^f \frac{\prod\limits_{m=p}^{n} k_m}{1 + k\beta} \frac{1 + k\beta}{k} = \frac{U_p^f}{\prod\limits_{m=1}^{p-1} k_m}. \quad (4.19)$$

Accordingly, whatever channel element contains the false signal source, the additive error does not depend on the feedback strength.

Therefore, multiplicative error suppression, considered in the previous section, cannot be applied to additive errors. The only way for error minimization is to find the error sources and to eliminate them.

In the electronics of the device, the main additive error source is the drive generator, producing a signal at the main frequency f, odd harmonics $(2n+1)f$ and also at even harmonics $2nf$ ($n = 1, 2, 3, \ldots$). Using the second harmonic of the EMF, even harmonics of the drive frequency cause a false output signal. They can enter the amplifier in different ways:

1. via the fluxgate sensor: The even harmonics from the drive are induced in the sense winding causing a magnetometer zero offset.

2. Even harmonics of the drive voltage can enter the amplifier via the power supply, and

3. even harmonics of the voltage can be coupled directly into the output amplifier.

A simple method for even harmonic detection in the magnetometer is using a switch to change the phase of the drive harmonics, and, therefore, the even harmonics phase by 180°. The synchronous detector shows these phase changes.

General methods for minimization and suppression of even harmonics are:

1. Selection of the operating condition such that nonlinear distortions are minimized;

2. the generator output design must be symmetrical as well as all other circuitry.

Any change in the drive signal phase results in a change of correlation between the inphase and quadrature parts and, hence, zero offset occurs. Frequency variations of the drive generator may cause strong zero offset.

The main frequency and the odd harmonics of the fluxgate at the amplifier input create (due to nonlinear characteristics) even harmonics of the output voltage, and, therefore, a false signal in the magnetometer output.

It is possible to detect this false signal in the magnetometer design, using switches at the input and output of the amplifier. The phase between the second harmonic voltage and the main frequency is unaffected by a 180° phase change of the main frequency (see Fig, 3.2). So, by synchronously switching at the amplifier input and output you can evaluate the second harmonic false signal level.

4.4. MEANS FOR ADDITIVE ERROR MINIMIZATION

Ways for reducing the voltage false even harmonics in the amplifier are similar to those for the drive generator: the selection of amplifier's operation conditions and the use of symmetrical output amplifiers.

The noise level on the magnetometer input of modern selective amplifiers is $0.5 - -2 \ pT$ at $0.001 - -1 \ Hz$, i.e. considerably lower than the fluxgate noise level.

Synchronous detectors and direct current amplifiers, including those operating as an integrator produce false signals of direct voltage. In the magnetometer these signals can be easily found by means of short-circuiting the synchronous detector input.

4.5 References chapter 1-4

1. V. K. Arkadyev, *Electro-magnetic Processes in Metals*. Part I (Russian). ONTI, Moscow, Leningrad (1934).

2. Yu. V. Afanassiev, V. P. Lulick, G. D. Alekseeva, Magnetometric Instruments of Space Stations "Luna–10" and "Venera–4" (Russian). *Cosmicheskiye issledovania (Space Research)* **6** (5), 772–781 (1968).

3. Yu. V. Afanassiev, *Fluxgates* (Russian). Energia, Leningrad (1969).

4. Yu. V. Afanassiev, N. V. Studentsov, A. P. Shyolkin, *Magnetometric Transducers, Instruments, Plants* (Russian). Energia, Leningrad (1972).

5. R. Ya. Berkman, R. E. Martynuk-Lototsky, Yu. I. Spector, Peculiarities of Ring Core Fluxgates' Calculation (Russian). *Automatichesky control i izmeritelnaya technica (Automation Control and Measurement Techniques)* **8**, 95–99 (1964).

6. R. Ya. Berkman, R. E. Martynuk-Lototsky, Some Peculiarities of the Physical Picture of Magnetic Modulators', Ring Core and Rod Core Fluxgates' Operation (Russian). In *Industrial Automation Magnetic Elements*, 107–112. Nauka, Moscow (1966).

7. L. L. Bernstain, About a New Magnetometer Type (Russian). *Izvestia AN SSSR (News of the Academy of Science of the USSR, Physics Series)* **8** (4), 189–193 (1944).

8. R. M. Bozorth, *Ferromagnetism*. D. van Nostrand Company Inc., Toronto, New York, London (1951).

9. S. I. Bondarenko, V. I. Sheremet, *Superconduction Application at Magnetic Measurements* (Russian). Energoatomizdat, Leningrad (1982),

10. G. I. Bochkov, Yu. E. Kuzovlev, News at $1/f$-Noise-study (Russian). *UFN (Physics Science Successes)* **141** (1), 151–176 (1983).

11. W. F. Braun (Jr.), *Micromagnetics*. Interscience Publishers – A Division of John Wiley & Sons, New York, London (1963).

12. L. L. Vanyan, *Electromagnetic Probing Basement* (Russian). Nedra, Moscow (1965).

4.5. REFERENCES CHAPTER 1-4

13. S. V. Vonsovsky, Ya. S. Shure, *Ferromagnetism* (Russian). Gostechisdat, Moscow, Leningrad (1948).

14. S. V. Vonsovsky, *Magnetism* (Russian). Nauka, Moscow (1971).

15. G. S. Gorelik, About Some Non-linear Phenomena Appearing at Interperpendicular Magnetic Field Superposition (Russian). *Izvestia AN SSSR (News of the Academy of Science of the USSR, Physics Series)* **8** (4), 172–188 (1944).

16. A. A. Grachev, About Spatial Correlation of Cycle Magnetization Reversal Noises (Russian), *DAN SSSR (Reports of the Academy of Science of the USSR)* **85** (4), 741–744 (1952).

17. P. Dial, K. Parkin, Moon's Magnetism (Russian), *UFN (Physics Science Successes)* **108** (1), 177–191 (1972).

18. Sh. Sh. Dolginov, L. N. Zhuzgov, V. A. Selutin, Magnetometric Instruments of the Third Soviet Satellite (Russian). In *Iskusstvennye sputniky Zemly (Earth's Satellites)* **4**, 135–160. Izdatelstvo AN SSSR (Printhouse of the Academy of Science of the USSR), Moscow (1960).

19. Sh. Sh. Dolginov, N. V. Pushkov, Study of Magnetic Fields of the Earth and Planets (Russian). In *USSR Successes in Space Research. First 10 years in Space: 1957–1967*, 177–249. Nauka, Moscow, (1968).

20. I. M. Kirko, Physical Similarity and Ferromagnetic Bodies Magnetization Analogy (Russian). *Izdatestwo AN Latv. SSR (Academy of Science Printhouse)*, Riga (1955).

21. N. N. Kolachievsky, *Magnetic Noise* (Russian). Nauka, Moscow (1971).

22. E. N. Langvagen, Methods for Determining Ferromagnetics' Differential Magnetic Permeability at Complex Magnetization (Russian). In *Izvestia vuzov (College News, Electromechanics)* **3**, 250–256 (1970).

23. Y. W. Lee, T. P. Cheatham, J. B. Wiesner, Application of Correlation Analysis to the Detection of Periodic Signals in Noise. *PJRE* **38**, 1165 (1950).

24. A. A. Logachev, *Magneto-searching Course* (Russian). Gostoptechizdat, Leningrad (1962).

25. L. M. Mandelshtam, *Lections on Oscillation Theory* (Russian). Nauka, Moscow (1972).

26. G. Marie, Amplificateurs paramétriques basses fréquences a commande magnétique orthogonale (French). *Acta Electronica* **8** (1), 2. Paris (1964).

27. E. A. Melnikov, Wide-band Ferrite Magneto-Modulation Sensor (Russian). *Geomagnetism i Aironomia* **1**, 183–185 (1975).

28. L. Ya. Mizuk, *Input Transducers for Low-frequency Magnetic Field Intensity Measurements* (Russian). Naukova Dumka, Kiev (1964).

29. D. D. Mishin, *Magnetic Materials* (Russian). Vysshaya shkola, Moscow (1981).

30. I. D. Morochov, E. P. Velichov, Yu. M. Volkov, Impulse MGD-generators and Deep Electro-magnetic Probing of the Earth's Rind (Russian). *Atomnaya energia (Nuclear Energy)* **41** (3), 213–219 (1978).

31. P. V. Novitsky, V. G. Knorring, V. S. Gutnikov, *Digital Instruments with Frequency Sensors* (Russian). Energia, Leningrad (1970).

32. B. V. Molotilov (Ed.), *Accuracy Alloys Reference Book*. (Russian). Metallurgia, Moscow (1974).

33. I. M. Puzey, B. V. Molotilov, Nickel-Iron-Molybdenum Alloys' Magnetostriction (Russian). *Izvestia AN SSSR (News of the Academy of Science of the USSR, Physics Series)* **22** (10), 1244–1250 (1958).

34. M. A. Rosenblat, Demagnetization Coefficients of High Permeability Rods. *J. Theoret. Phys.* **24** (4), 637–661 (1954).

35. M. A. Rosenblat, To Calculation of Magnetic Field Intensity Magneto-Modulation Sensors (Russian). *Electrichestvo (Electricity)*, 24–31 (1957).

36. M. A. Rosenblat, *Magnetic Elements of Automation and Hardware* (Russian). Nauka, Moscow (1966).

37. A. G. Samoylovich, L. L. Korenblit, Thermo-electric Eddy Currents in Anisotropic Media (Russian). *Physica twerdogo tela (Solid State Physics)* **3** (7), 2054–2059 (1961).

38. N. M. Semenov, N. I. Yakovlev, *Digital Fluxgate Magnetometers* (Russian). Energia, Leningrad, (1978).

39. V. V. Kolachevskaya, N. N. Kolachevsky, V. V. Rozhdestvensky, L. V. Strygin, Spectral Distribution of Magnetic Noise Around Magnetization Reversal Frequency Harmonics (Russian). *Radiotechnica i Electronika* **16** (7), 1211–1215 (1971).

40. Yu. V. Afanassiev, N. V. Studentsov, V. N. Horev, E. N. Chechurina, A. P. Shelkin, *Means for the Measurement of Magnetic Field Parameters* (Russian). Energia, Leningrad (1979).

41. E. Philippow, *Nichtlineare Elektrotechnik* (German). Akademische Verlagsgesellschaft Geest und Partig KG, Leipzig (1963).

42. A. A. Harkevich, *Nonlinear and Parametric Phenomena in Radiotechnics* (Russian). GITTL, Moscow (1956).

43. V. Ya. Shifrin, *Instruments for Precise Measurements of the Induction Density and its Metrological Provision* (Russian). Mashinostroyeniye, Moscow (1982).

44. G. A. Stamberger, K. B. Karandeev (Ed.), *Devices for Weak Constant Magnetic Field Formation* (Russian). Nauka, Novosibirsk (1972).

45. S. Ya. Yavor, M. Siladyi, Simultaneous Magnetic Field Formation by Rectangular Solenoid of Finite Length (Russian). *PTE* **1**, 147–149 (1961).

46. N. I. Yakovlev, Peculiarities of Fluxgate Sensor Operation at Resonance Mode. *Geophysicheskaya apparatura* **35**, 27–28 (1968).

47. E. P. Felch et al., Air-Borne magnetometer for search and Survey. *AIEE Trans.* **66**, 641–651 (1947).

48. M. H. Acuña, Fluxgate magnetometers for outer planet exploration. *IEEE Trans. on Magn.* **10** (3), 519–523 (1974).

49. M. H. Acuña, *MAGSAT vector magnetometer – Absolute Sensor alignment determination*. NASA Tech. Memo., 79648 (1981).

50. M. H. Acuña, The UOSAT magnetometer experiment. *The Radio and Electronic Engineer* **52** (819), 431–436 (1982).

51. H. Aschenbrenner, G. Goubau, Eine Anordnung zur Registrierung rascher magnetischer Störungen (German). *Hochfrequenz und Elektroakustik (Jahrbuch der drahtlosen Telegraphie und Telephonie)* **47** (6), 177–181 (1936).

52. H. Bittel, Noise of ferromagnetic. *IEEE Trans. on Magn.* **5** (3), 359–365 (1969).

53. J. R. Burger, The Theoretical Output of Ring Core Fluxgate Sensor. *IEEE Trans. on Magn.* **8** (4), 791–798 (1972).

54. P. Dyal, D. I. Gordon, Lunar Surface Magnetometers. *IEEE Trans. on Magn.* **9** (3), 226–231 (1973).

55. M. Candidi, R. Orfei, F. Palutan, G. Vannaroni, FFT analysis of a space magnetometer noise. *IEEE Geosci. Electron.* **12**, 23–28 (1974).

56. W. A. Geyger, New Type of flux-gate Magnetometer. *J. Appl. Phys.* **33** (3), 1280–1281 (1962).

57. D. I. Gordon, R. H. Lundsten, R. A. Chiardo, Factors Affecting the Sensitivity of Gamma-Level Ring Core Magnetometers. *IEEE Trans. on Magn.* **1** (4), 330–337 (1965).

58. D. I. Gordon, R. H. Lundsten, I. E. Scarzello, Offset and Noise in Fluxgate Magnetometers. *IEEE Trans. on Magn.* **6** (4), 818 (1970).

59. D. I. Gordon, R. E. Brown, Recent Advances in Fluxgate Magnetometry. *IEEE Trans. on Magn.* **8** (1), 76–83 (1972).

60. G. Gorelik, X. Goronina, I. Joukova, Sur les courbes d'aimantation longitudinale d'un fil ferromagnétique parcouru par un courant continu (French). *Compt. Rend. Acad. Sci. (USSR)* **44**, 235–237 (1944).

61. F. Mobley, L. D. Eckard, C. H. Fonntain, C. W. Ousley, Magsat – a new satellite to survey the earth's magnetic field. *IEEE Trans. on Magn.* **16** (5), 758–760 (1980).

62. N. F. Ness, *Magnetometer for space research.* Goddard Space Flight Center, Rome (1970).

63. F. Primdahl, The fluxgate mechanism. *IEEE Trans. on Magn.* **6** (1), 376–383 (1970).

64. F. Primdahl, The fluxgate magnetometer. *J. Phys. E.: Sci. Instrum.* **12** (4), 241–253 (1979).

65. F. Primdahl, P. Anker-Jensen, Compact spherical coil for fluxgate magnetometer vector feed back. *J. Phys. E.: Sci. Instrum.* **15**, 221–226 (1982).

4.5. REFERENCES CHAPTER 1-4

66. G. L. Romani, S. J. Williamson, L. Kaufman, Biomagnetic instrumentation. *Rev. Sci. Instr.* **53** (12), 1815–1815 (1982).

67. C. T. Russell, The ISEE 1 and 2 Fluxgate Magnetometers. *IEEE Trans. on Geo. Electr.* **16** (3), 239–242 (1978).

68. R. D. Russell, B. B. Narod, F. Kollar, Characteristics of the Capacitively Loaded Flux Gate sensor. *IEEE Trans. on Magn.* **19** (2), 126–130 (1983).

69. D. C. Scouten, Sensor noise in low-level fluxgate magnetometer. *IEEE Trans. on Magn.* **8**, 223–231 (1972).

70. P. H. Serson, S. Z. Mack, K. Witman, A threecomponent airborne magnetometer. *Publ. Dom. Obs.* **19** (2), 15–97 (1957).

71. P. H. Serson, L. W. Hannaford, A portable Electrical Magnetometer. *Canad. J. Tech.* **34**, 232–243 (1956).

72. K. Shirae, Noise in amarphus magnetic materials. A Publication of the IEEE Antennas and Propagation Society, Vol. AP-02 (1984), ISSN CH 1918-2/84. *IEEE Trans. on Magn.* **20** (5), 1299–1301 (1984).

73. R. C. Snare, R. L. McPherron, Measurement of instrument noise spectra at frequencies below 1 hertz. *IEEE Trans. on Magn.* **9**, 232–235 (1973).

74. Z. Storm, Ch. Heiden, Untersuchungen über das Frequenzspektrum des Barkhausen-Rauschens (German). *Z. Angew. Phys.* **17** (3), 161–164 (1961).

75. V. V. Vacquir, R. F. Simons, A. W. Hull, A Magnetic Airborne Detector Employing Magnetically Controlled Gyroscopic Stabilization. *RSI* **18** (7), 483–487 (1947).

76. M. M. Weiner, Magnetostrictive offset and noise in fluxgate magnetometers. *IEEE Trans. on Magn.* **5**, 98–105 (1969).

77. M. Wurm, Beiträge zur Theorie und Praxis des Feldstärkedifferenzmessers für magnetische Felder nach Förster (German). *Z. Angew. Phys.* **2** (5), 210–219 (1950).

78. E. Keppler, E. Kirsch, P. Mörl, G. Musmann, A. Rossbach, L. Rossberg, Raketenexperiment zur Untersuchung von Nordlichtern (German). *Z. Geophys.* **33**, 346–361 (1967).

Part II - Applications

Chapter 5

General Fluxgate Magnetometer Design

contributions from

M.H.Acuña[1], Y.V.Afanassiev[2], A.Balogh[3], C.Carr[3], F.Kuhnke[4],
G.Musmann[4], F.Primdahl[5] and R.C.Snare[6]

[1] NASA, Goddard Space Flight Centre, Greenbelt, Washington
[2] MAG − Sensors, St.Petersburg
[3] The Blacket Laboratory, Imperial College London
[4] formerly Institute of Geophysics, Technical University Braunschweig
[5] Danish Space Research Institute, Lyngby, Copenhagen
[6] UCLA, Institute of Geophysics and Planetary Physics, Los Angeles

5.1 Instrument Design

5.1.1 General

This chapter is a compilation of reports, papers and articles about Fluxgate Magnetometers applied in Space written by the authors listed above in the time period 1970-2002.

Fluxgate magnetometers have been widely used for all kind of magnetic field investigations in space like accurate Earth Field measurements, the study of Planetary magnetism, Interplanetary magnetic fields and Solar Wind interactions with celestrial bodies. These missions require precise vector magnetic measurements with wide dynamic range, high stability in time and versus temperature, high reliability, low noise, low and stable zero levels or offsets, high and stable linearity. From the earliest spacecrafts to those of today fluxgate magnetometers have been used due to their inherent simplicity, low power consumption , vector response characteristics and low mass. To make the precise vector magnetic measurements in space required by the scientific investigation one needs a fluxgate magnetometer system whose elements include the fluxgate vector sensor, the analog electronics, the analog-to-digital converter and the data system. During the last about forty years from the earliest spacecraft to those of today there has been a constant increase in science requirements based on the availability of greater spacecraft resources including telemetry bandwidth , power and mass but also the advances in high reliability and red hard parts as well as in electronics such as low power analog and digital integrated circuits and microprocessor systems.

5.1.2 The Ring Core Fluxgate

The majority of vector measurements in space have been made with fluxgate sensors. Aschenbrenner and Goubau developed in 1936 the first fluxgate sensors. Since those days the principle of operation has not changed. An alternating drive current in the sensor drive winding drives the permeable core material alternately deep into saturation. Because of the non-linear coupling due to core saturation the induced voltage in the sense winding has all kind of harmonics. The amplitude of the even harmonics is proportional to the ambient magnetic field aligned with the sense winding. Normally the second harmonic is filtered out and synchronously detected to produce a voltage proportional to the magnetic field. In most cases the output voltage is fed back through a scaling resistor to the feedback winding on the sensor. As a result you get a highly linear, stable vector measuring instrument. A number of magnetic core configurations have been used since this early development. The configuration that was used in the early years of space exploration was the parallel core design developed by Förster in (1937), subtracting the primary drive frequency in the pick-up winding around the two anti-parallel driven cores (Explorer 10,12 and 14, Imp1 and 4, ATS1, Pioneer 6 Mariner 2,Electron 2,Lunik 1,2 and 10 and Sputnik 3 ,German AZUR and Helios A and B spacecrafts).This core configuration however requires high drive power. Cylindrical core magnetometers have been constructed by Schonstedt (1959) and Primdahl(1970).The Schonstedt sensors consist of a ceramic cylinder with thin Permalloy tape helically wound around the cylinder, thus the name HELIFLUX sensor. The drive winding is applied toroidally through

5.1. INSTRUMENT DESIGN

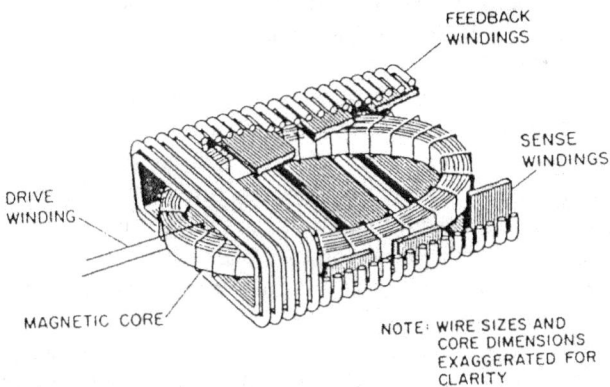

Figure 5.1: Ring core sensor(Robert C.Snare,UCLA : A History of Vector Magnetometry in Space)

the tube with a solenoid feedback coil around the tube. This HELIFLUX sensor resulted in low noise and stable offsets , it was used on a number of space flights. The ring core magnetometer was first used by Aschenbrenner and Goubau(1936), revived by Geyger (1962) and refinements were made at Naval Ordonance Laboratory, White Oak. These modern ring core uses several layers of 0,0254mm Permalloy wound around a non magnetic stainless steel ring while the excitation winding is toroidally wound around the ring. Sense and feedback windings are wound on a bobbin on the outside of the ring defining the sense axis of the sensor as shown in Figure 5.1(Both noise and offset performance was enhanced with the development of 6-81.3 Mo-Permalloy a low magnetostriction tape material by Gordon et al.(1968).

These cores were first used in space in the Lunar Surface Magnetometer package ,developed by Naval Ordonance Laboratory, White Oak,(Dyal and Gordon,1973), left on the moon by the Apollo 16 astronauts in 1972. Today nearly all vector scientific measurements in space besides the Vector Helium Magnetometer are made with ring core magnetometers following the design proposed by D.Gordon. The first fluxgate magnetometer was carried into Earth orbit May 1958 aboard the Soviet artificial Earth satellite Sputnik 3, Dolginov et al.(1960). The first vector magnetic measurements in space were made aboard Luna 1 and 2 (Dolginov et al.,1961). This core is made by winding a thin Permalloy ,82NiMo tape of 0,02mm gauge and 1,5mm width , polished ,degreased and coated with a heat resistant insulation over a metallic bobbin and annealing them together. The core with the frame is placed inside the flat pick-up coil such that it may be rotated around the ring center for tuning offset and noise. The bobbin is made from non-magnetic resistive alloy NiMoCrAl having a thermal expansion coefficient close to the Permalloy coefficient $12^{-6}K^{-1}$ and melting temperature ($1350°C$) The bobbin in form of a thin ring is polished and iron free. The six turns of the Permalloy tape are wound into the groove with constant tension where the end is being fixed to the preceding turn one by spot welding. The core is then annealed to the bobbin at $800 - 1000°C$ with subsequent slow cooling ($100°C/h$) in the temperature range $600 - 350°C$. The core dimensions are selected ac. to ch.2 and 3 of part I, while the Permalloy tape gauge of 0,02mm is taken in accordance with the expression (3.5) for the excitation frequency f1=12,5kHz (see sect.5.1.3).The core diameter of 13.2mm was chosen to provide a reliable spatial correlation of noise along the core(sect.3.4), to exclude the magnetostriction resonance (see sect.3.6), to achieve a small design and still have an acceptable conversion coefficient in combination with the following amplifier.The permeability of the ring core (with respect to the measured field) is estimated by the formula (1.53, part I) and taking into account that the total core thickness $\rho = 6 \cdot 0.02 = 0,12$ mm constitutes the value m=416. A single layer toroidal excitation winding, which has 130turns, covers the core uniformly.The maximum value of excitation current flowing through this winding is equal to 0.6-0.8A,and creates in the core an excitation field of 2-2.5kA/m.

Taking into account the advantages and possibilities of modes, providing the shape of wave of the excitation field to be close to rectangular (see Sect. 2.6), the estimation of the fluxgate conversion coefficient in an off load mode is done in accordance with the formula (2.66). Under $f_1 = \omega/(2\pi) = 12,5$ kHz, $s_\Sigma = 2\rho\delta = 0.36$ mm^2, $\mathcal{W}_2 = 500$ (the number of turns in the sense coil), $\xi_1 = \xi_2 \approx 0.6$ and $m = 416$ one has $G_{2\max} \approx 4.5$ μV/nT. This value is in a good agreement with the measured one (≈ 5 μV/nT).

The fluxgate tuning is performed by rotation of the core together with drive coil with respect to the stationary sense coil. In the best specimens the misbalance coefficient on the second harmonic is $\varepsilon = (2\,to-5)10^{-3}$, means the second harmonic false signal, produced by the generator, is reduced by 200–500 times. In this case the levels of the first and third harmonics of the EMF of misbalance do not exceed 10 mV. The tuning is being done during construction of the fluxgate (see Sect5.1) in accordance with the procedure, described in Sect. 3.2.

It was found that the false signal of the second harmonic penetrates into

5.1. INSTRUMENT DESIGN

the sense coil due to an asymmetrical capacitance relationship between the sense and drive coils. In order to exclude this possibility the sense coil is made from two sections, which are wound up in such a way that the first and last layers of turns, as well as their output ends, are at the same distance from the drive coil.

The fluxgate contains a compensation coil of a rectangular form, and also a number of auxiliary windings, which are wound above it, such as a calibration winding, a correction one, etc. The constant of the compensation coil is estimated by the following formula:

$$C_{\mathrm{K}} = \frac{\mu_0 \mathcal{W}}{2\pi l} \left(\arctan \frac{l+x}{ab}\sqrt{(l+x)^2 + a^2 + b^2} \right.$$
$$\left. + \arctan \frac{l-x}{ab}\sqrt{(l-x)^2 + a^2 + b^2} \right), \qquad (5.1)$$

where μ_0 is the magnetic constant, $\mathcal{W}$ is the number of turns of the coil, l, a, b are half values of the length, the thickness, and the width of the coil, x is the current coordinate along the longitudinal axis of the coil from its center. When the geometric parameters of the coil are equal (in millimeters) to $l = 15$, $a = 6$, $b = 10$ and the number of turns satisfies $\mathcal{W} = 200$, the constant in the coil center is equal to $C_{\mathrm{K}} = 7.310^{-3}$ T/A = 7.3 nT/μA. The constant is reduced by 2.5 % at a distance $x = 5$ mm from the coil center.

All the fluxgate components, except the core and the bobbin, are pressed from the DSV(=wooden chip fibre, Russian name) materials and AG-4V(=glass reinforced plastics), which have a linear thermal expansion coefficient of the order $10^{-5} K^{-1}$. The coils which are wound on the pressed frames are covered with a cementing compound. Acuña et al. in (1978) selected the highly stable glass ceramic material MACOR , having an expansion coefficient of $10^{-5} K^{-1}$ for the very precise MAGSAT magnetometer structure elements (see chap.7,fig7.2).Its feedback coil was wound with polyurethane-nylon insulated platinum wire with a linear thermal expansion coefficient of $9,9 \times 10^{-6} K^{-1}$. The ring core material was 6-81 Mo Permalloy.

The following are the characteristics of the fluxgate for an excitation current, of frequency $f = 12.5$ kHz and amplitude 0.6 A in an impulse (obtained by testing over 100 specimens):

- the conversion coefficient on the frequency $2f$ in the idle mode equals to $G_{2\mathrm{i.m.}} \approx 5~\mu\mathrm{V/nT}$;

- the conversion coefficient in the mode of tuning of the sense coil to the resonance for the frequency $2f$ is $G_{2\mathrm{r.}} \approx 40~\mu\mathrm{V/nT}$;

142 CHAPTER 5. GENERAL FLUXGATE MAGNETOMETER DESIGN

- the misbalance EMF on the frequencies f and $3f$ in the idle mode are equal to E_{m1} and $E_{m3} < 10$ mV;

- the misbalance coefficient on the frequency $2f_{\varepsilon 2} < 10^{-2}$;

- the admissible zero offset $B_0^m < 1$ nT;

- the instability of zero during 5–10 days under normal conditions is $B_0^m(t) \leq 0.2$ nT;

- the noise level in the range 0.01–1 Hz is $\overline{B}_n \approx 12$ pT

- the noise spectral density in the frequency above 1 Hz is $b_n < 5$ pT Hz$^{-1/2}$.

Somewhat better results with respect to the noise ($\overline{B}_n = 7\text{--}9$ pT and $b_n \leq 3$ pT Hz$^{-1/2}$) are obtained by replacement the alloy 82NiMo by the alloy 83NiV(Institute for Precision Alloys,Russia). However practically same results were obtained for the alloy 82NiMo after an increase in the drive current to 0.9–1 A. In conclusion,it is stressed, that the ring type fluxgate in compari-

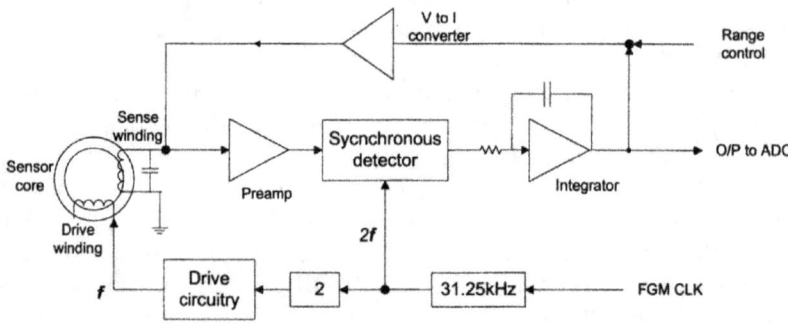

Figure 5.2: Simplified single axis fluxgate magnetometer sensor electronics block diagram(C.Carr et al.: Double Star Magnetic Field Investigation)

son with other fluxgates containing closed cores (for example of the race-track type, see Sect. 2.5) do have the best characteristics with respect to noise and zero stability .This is why their usage in instruments with low sensitivity threshold is reasonable not only nowadays, but also in the future, when ferromagnetic substances with less intrinsic noise are found (see Sect. 3.7). Acuña (1970) describes two ring core fluxgate magnetometer concepts , one for high field measurements with an upper limit capability of more than

5.1. INSTRUMENT DESIGN

2×10^6 nT and a second one for low field measurements with sensitivities approaching $0.01 nT$ in $0-10Hz$ bandwidth and zero level stability of the order of $+/-0.1nT$. These designs have been used for Pioneer 11 and MJS 77 magnetic field instrumentation. The Pioneer 11 fluxgate magnetometer design features extreme simplicity and low power consumption. The sensors consist of two orthogonally mounted ring cores, driven cyclically to saturation by their associated oscillators at a frequency of 8kHz. Two orthogonal windings diametrically wound around each core, provide sensitivity along three orthogonal axes plus a redundant measurement along one of the axes. A distinctive feature of the design is that is does not operate on the second harmonic principle in a strict sense, but rather derives its output from the short pulses induced in the sense windings when an external magnetic field is present. The D.C. output is obtained by phase sensitive detection of the reference and signal pulses. Two factors account for the nonlinear response characteristics for large external fields: a) the attenuation factor is a non-linear function of the applied field, and b) the drive frequency is field dependent and tends to increase with increasing external fields as the flux traverses a smaller hysteresis loop. Since no tuning of the sense windings is required, these effects can be used to advantage to obtain a quasi logarithmic response. Thus it was possible to cover a large dynamic range in field intensities with a limited output voltage range. The high field sensor are constructed with 1cm diameter 4-79 Molybdenum Permalloy cores while the low field sensors are constructed using low noise, high stability, 2.54cm diameter ring cores developed by Gordon as shown above.

5.1.3 The Drive Generator

As it was noted in the Sect. 4.4, the excitation square wave generator should have a high frequency stability and a low level of even harmonics in the output voltage. These requirements are met using typically a crystal clock of about 60kHz, first dividing its frequency by a factor of 2 for the synchronous detector and again by a factor of 2 for the ring core current drive. Fig. 5.2 shows a typical block diagram of a single axis fluxgate magnetometer including the sensor, the drive generator (FGM clock), the preamplifier, synchronous detector, integrator, feedback amplifier and signal output to the analog-digital converter. This is the state of the art block diagram as shown by C. Carr et. al for the Double Star Magnetic Field Investigation.

It should be noted that it is always desirable to drive the core as deep into saturation as possible to reduce perming effects and noise while maintaining low power consumption. An efficient arrangement used, first in the MJS (Mariner Jupiter Saturn)design to obtain large peak drive currents with low average power consumption is described by Geyger and other authors.As the core traverses the high permeability region of the B-H curve, the impedance reflected in parallel with a capacitor is high and the capacitor charges through inductor, when the core reaches saturation, the reflected impedance decreases to a very small value and the capacitor essentially discharges across the winding. A large current pulse of short duration will flow in the winding, transferring the electrostatic energy stored in the capacitor into electromagnetic energy in the small saturated inductance of the drive winding.Then this energy is returned to the capacitor reversing its voltage.It is clear that the arrangement constitutes a non-linear oscillating circuit with its natural frequency determined by the core saturation flux capacity, number of turns in the winding and the initial capacitor voltage. With appropriate component values the ratio of peak drive to average drive current can approach large values e.g. 50 (on MAGSAT it was e.g. 100 times the coercive force)and high efficiency is therefore obtained.When low power consumption is not a major requirement, the inductor may be replaced by a resistor with analogous results. A significant advantage of this circuit is that the current waveform in the drive winding is essentially independent of the drive voltage waveform as long as the latter possesses odd symmetry and sufficient amplitude.This feature eliminates the requirement for low distortion, amplitude stabilized, sinusoidal voltage sources and allows the use of simple square wave circuits as shown in the block diagram above.

This approach has been implemented in most ring core fluxgate magnetometers since those days of MJS.

A problem with fluxgate magnetometers is the temperature dependence of the output voltage.Using state of the art electronic components and employing dynamic feedback technique it is possible to eliminate the influence of temperature changes on the electronics. The sensors however, still cause temperature changes of a few nT per $°C$ in high fields like Earth Field.These temperature dependence are proportional to the field being measured.According to Primdahl the source of drift is believed to be the bucking field created by the feedback current in the secondary coil.Primdahl (1970) describes a temperature compensation of fluxgate magnetometers in details. Kuhnke (private communication)describes a shorter version of how to control the mechanical stability of the feedback coil and the stability of the current passing through it.To keep the mechanical alterations as small as possible he chooses the glass ceramic MACOR for the bobbin and the whole support structure of the sensor head.This ultra small design was integrated on NASA's DEEP SPACE ONE mission 1998 by D.E.Brinza, M.D.Henry (NASA-JPL, see M.D.Henry et al.),F.Kuhnke and G.Musmann(for measuring the plasma turbulence around the first(Xenon) Ion engine spacecraft propulsion beam and

5.1. INSTRUMENT DESIGN

comet Borelly magnetic fields and plasma waves)and on ESA's cometary mission ROSETTA . The dual ring core sensor design ,one perpendicular inside the other one,thus having a centered zero point ,was only $25 \times 25 \times 25 \times mm^3$ with a mass of 28g and a power consumption of 200mW. This sensor can be operated in a temperature range from $-150°C$ to $+300°C$.

The design was extensively tested on a high temperature ($300°C$) downhole(deep drilling) probe magnetometer using a similar mechanical sensor design. The variation of the measured field component is mainly dependent upon the variation of the coil length respectively the copper wire resistivity change. To get rid of the temperature dependency it is possible to benefit from the strongly temperature dependant resistivity of the copper wired sensor coils,which means the coil resistances are used as senors for each coil temperature.To keep the coil's current independent from its resistance it is necessary to provide a device which is able to generate an appropriate negative resistor. This can be achieved with an operational amplifier ,wired as a voltage controlled current source with grounded output.This is similar to a negative impedance converter and creates negative resistors. For stability net negative feedback must be maintained around the op amp.to get the feedback field (current) independent of temperature only one resistor acc.to Kuhnke(private communication) has to be adjusted (second order terms omitted). (For details see References chapter 5, F.Kuhnke et al. 1998) Some ring-core magnetometers that are used in 'strong'background magnetic fields often show the effect of **non linear cross-coupling** this is discussed by Acuña 1978 and in chapter 9. .

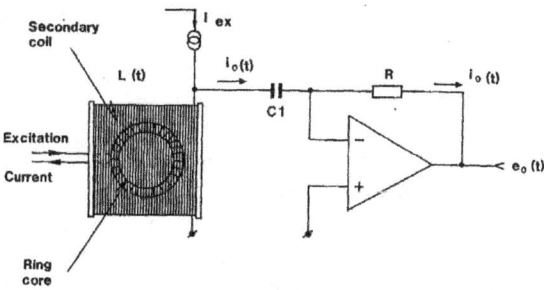

Figure 5.3: Short-circuit input stage converting the coil current $i_0(t)$ to output voltage $e_0(t) = R \cdot i_0(t)$. A DC-current I_{ex} will compensate for the external magnetic field B_{ex}

5.1.4 The Short-Circuited Fluxgate Sensor

Introduction

The fluxgate sensor is most commonly used in the open-circuit mode by coupling the unloaded secondary pickup coil to the high impedance input of the conditioning amplifier, or in the capacitively loaded mode, where the sensor output is tuned to the second harmonic of the excitation frequency in order to enhance the second harmonic signal and attenuate the odd harmonics feed-through signals. Many theoretical treatments deal with the unloaded case (see Primdahl, 1979, and references therein); but also the second harmonic tuned fluxgate has been treated quite extensively starting with the classical paper by Serson and Hannaford (1956) and the works by the Canadians (Russell, Narod and Kollar, 1983; Narod and Russell, 1984; Gao and Russell, 1987), and by Player (1988). The noise characteristics of the tuned fluxgate were treated by Primdahl and Jensen (1987).

However, the output current impulses from the short-circuited secondary coil contains as much information about the external magnetic field as does the open-circuit voltage, and the output current signal shows less 'ringing' because the cable and coil capacitances are short-circuited. The output current impulses are independent of even large changes in the excitation level beyond saturation, and the broad-band input amplifier combined with an all-even-harmonics time-domain detection system opens for the development of a greatly simplified and high performance analogue circuit without the need for second harmonic tuning (Primdahl et al., 1989).

The short-circuit principle was first successfully tested on board the NASA 35.024UE sounding rocket launched December 1988, and the Earth's field mapping CSC-magnetometer on board the Danish Ørsted satellite is based on this principle.

The Output Current of the Short-Circuited Fluxgate Sensor

Figure (5.3) shows the input amplifier stage, which short circuits the secondary pickup coil of a conventional ring core sensor (see e.g. Nielsen et al., 1995). The operational amplifier is preferably of the current feedback type, because this circuit has an inherently low input impedance which avoids the tendency of instability often encountered with voltage feedback types when used in this circuit.

In an external magnetic field B_{ex} a short-circuit current $i(t)$ will flow in the coil caused by the periodically changing self-induction $L(t)$ (Primdahl et al., 1989, 1991, Primdahl, 1992). Considering a frequency range where the

5.1. INSTRUMENT DESIGN

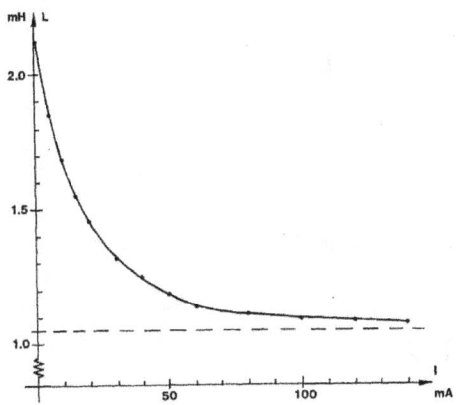

Figure 5.4: The output self-induction L of a 17 mm ring core sensor against the DC-current I in the core excitation winding

circuit losses can be ignored, we may write:

$$\Phi_{\text{coil}} = (I_{\text{ex}} + i(t)) \cdot L(t) = constant \quad (5.2)$$

Here $I_{\text{ex}} = (B_{\text{ex}}/\mu_0) \cdot (l/N)$, where (N/l) is the effective secondary coil winding density, and I_{ex} is equal to the (virtual) DC-current in the coil, that would exactly match and compensate the external field B_{ex} inside the coil. The coil is, of course, not superconducting, which means that the time-average of $< i(t) >= 0$, and we have:

$$\Phi_{\text{coil}} \cdot < 1/L(t) >= I_{\text{ex}} \quad (5.3)$$

Introducing the average self-induction $L_0 = 1/ < 1/L(t) >$ we get:

$$i(t) = I_{\text{ex}} \cdot \left(\frac{L_0}{L(t)} - 1 \right) \quad (5.4)$$

and this is the very simple basic equation relating the external field B_{ex} (via I_{ex}) to the short-circuit current $i(t)$ and to the time-varying fluxgate self-induction $L(t)$.

The peak-to-peak magnitude of the current impulses is:

$$\Delta(i_p - p) = I_{ex} \cdot \left(\frac{L_0}{L_{max}}\right) \cdot \left(\frac{L_{max}}{L_{min}} - 1\right) \tag{5.5}$$

where the factor $\left(\frac{L_{max}}{L_{min}} - 1\right)$ is a figure of merit for the fluxgate sensor, and for ringcores it typically ranges from 0.8 to 1.2 depending on the geometry. Approximately, we have
$L_{max} = L_{coil}$ (when $i_{exc} = 0$), and $L_{min} = L_{air}$ (without the core), which are both easily measured.

The fraction of time spent in saturation by the core is δ, and then the (inverse) average self-induction is given by:

$$\frac{1}{L_0} = \frac{\delta}{L_{max}} + \frac{1-\delta}{L_{min}} \tag{5.6}$$

Figure 5.4 shows the relation between the secondary pickup coil self-induction $L(i_{exc})$ and the core excitation current i_{exc} for a 17 mm diameter ring core (Primdahl et al., 1989). For increasing i_{exc} the self-induction drops rapidly from $L_{max} = 2.12 mH$, and approaches the air-cored self-induction $L_{air} = 1.05$ mH, when i_{exc} exceeds $50\,to\,100\ mA$.

Figure (5.5) shows the excitation current and the short-circuit output current impulses, when the sensor is exposed to the Earth's field of about 49,000 nT. The calculations fit the measured values within a few percent (Primdahl et al., 1989, Nielsen et al., 1995).

The maximum coil self-induction is expressed by:

$$L_{coil} = L_{air} + L_{air} \cdot (\mu_a - 1) \cdot \frac{A_{core}}{A_{coil}} \tag{5.7}$$

where A_{core} and A_{coil} are the core and coil cross sectional areas, respectively, and where the apparent permeability is given by:

$$\mu_a = \frac{\mu_r}{1 + D \cdot (mu_r - 1)} \tag{5.8}$$

By measuring L_{coil}, L_{air}, A_{core}, A_{coil} and μ_r it is then possible to calculate the effective demagnetizing factor D for any particular fluxgate sensor (Primdahl et al., 1989a). For typical ring cores D falls in the range from $10^{-3}\,to\,10^{-2}$. Nielsen et al.(1995)present a theoretical calculation of the ring core demagnetizing factor in excellent agreement with the measured values.

5.1. INSTRUMENT DESIGN

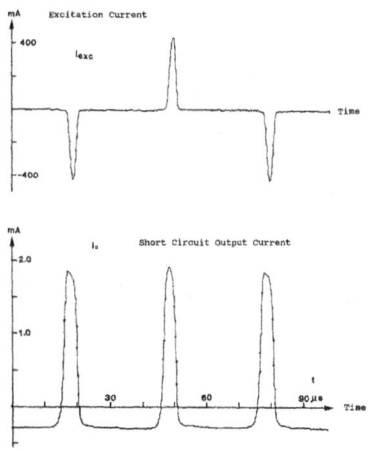

Figure 5.5: The ring core excitation current $i_{exc}(t)$ and the short-circuit output current impulses $i_0(t)$ for a 17 mm ring core exposed to an axial field of about 50,000 nT.

The All-Even-Harmonics Detection Feedback Magnetometer

Running the fluxgate sensor in short-circuited mode does not, of course, preclude the use of subsequent second harmonic narrow band pass filters and a conventional second harmonic phase detector in order to obtain a DC-signal proportional to the axial sensor field. However, the discrete output current impulses may be time-domain detected with a time-controlled switch connecting the impulses to the sampling and integrating output stage. Figure (5.6) shows the resulting very simple circuit for a single-component magnetometer. The switch control impulses must be phase adjusted relative to the excitation and adjusted in width to match the output current impulses. C_s is a sample-and-hold capacitor and the output stage is a two-pole low pass filter effectively attenuating the switching impulses. The feedback current flows through R_3, which, owing to the short-circuited mode, does not attenuate the sensor signal by loading the secondary coil. The open loop noise and drift performance of the circuit depends mostly on the quality of the switching circuit, and it matches or surpasses the best second harmonic detection systems. A detailed discussion of the performance of this circuit can be found in Nielsen et al. (1995) describing the development and characteristics of the

150 CHAPTER 5. GENERAL FLUXGATE MAGNETOMETER DESIGN

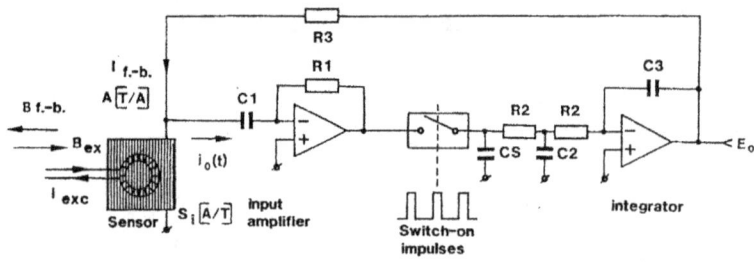

Figure 5.6: Single component magnetometer with short-circuited sensor

CSC-fluxgate magnetometer.

According to the simplified block diagram Fig 5.2 the following signal processing electronics consists of a low noise FET A.C.preamplifier tuned to the second harmonics of the drive signal providing a gain of approximately 75, a synchronous demodulator, an operational integrator and a transconductance feedback summing amplifier instead of a feedback resistor in conventional designs.A low pass antialiasing filter is utilized at the output to limit the signal bandwidth depending on the following analog-digital converter sampling rate. On MAGSAT for example the overall sensitivity achieved was

$$100 \mu V/nT$$

at the secondary of the input transformer(compared to

$$10 mV/nT$$

for copper wound coils). In the USSR in the beginning of 1960 the magnetometers SG-52, SG-55[1] and SG-56 were designed, and later more sensitive ones were developed, SG-59M and SG-70, having a range for each component equal to ±50 nT, instability of zero during 10 days being not more than 3 nT and a sensitivity threshold equal to 0.1 nT. The schematic diagram and

---

[1]SG means Geophysics Station

5.1. INSTRUMENT DESIGN

the design of the magnetometer SG-59M is described in details in the works [2, 3]. The magnetometer was installed on the space station "Luna-10" which was the first Moon satellite. It was with these magnetometers, with some modifications and some devices for automatically expanding their measurement range, that the basic evidence was collected concerning the magnetic fields of the Moon, Venus and Mars and the interplanetary magnetic field, these investigations were generalized in the works [19].

Similar purpose magnetometers were also designed in the USA [62]. As a rule, the Americans employed closed core fluxgates, using for the cores, among others, some new low noise ferro-nickel alloys [54, 57, 58, 59], that let them not only to reduce the threshold of sensitivity by one order, but also to improve temporal and thermal stability of the instruments.

References chapter 5

Henry,M.D.,D.E.Brinza,A.T.Mactutis,K.P.McCarty,J.D.Rademacher,T.R.van Zandt, R.Johnson,S.Moses,G.Musmann and F.Kuhnke, NSTAR Diagnostic Package(NDP) Architecture and Deep Space One (DS1) Spacecraft Event detection,NASA-JPL,IEEE Aerospace Conf.,Big Sky,MT.,Calif.,2000. (JPL TRS 1992+)

Kuhnke,F.,M.Menvielle,G.Musmann,J.F.Karczewski,H.Kügler,C.Cavoit and P.Schibler, The OPTIMISM/MAG Mars-96 experiment: magnetic measurements onboard landers and related magnetic cleanliness program, Planet.Space SCI.,Vol.46,No.6/7,pp.749-767,1998

All other References are already included in chapter 12, the General Fluxgate Magnetometer Bibliography.

Chapter 6

Magnetic Materials for Sensors

O.V.Nielsen[1], Y.V.Afanassiev[2] and K.H.Fornaçon

[1]Dept.of Electrophysics, Technical University of Lyngby,
[2]MAG − Sensors, St.Petersburg,
[3]Max − Planck − Institut für Extraterrestrische Physik, Adlershof/Berlin

6.1 Common Requirements for the Magnetic Materials

Ferromagnetic materials may be divided in magneto-soft and magneto-sharp materials according to the shape of the hysteresis curve. A criterion for this division is the value of the coercivity H_C which is simple for measuring. Then, magneto-soft materials generally have relatively low coercivities whereas magneto-sharp materials have higher coercivities. As a rule magneto-soft materials usually are used in fluxgate sensors. For crystalline alloys the best results are obtained with extreme magneto-soft materials, see Sect. 6.2.

Potential magnetic materials are:

- crystalline nickel rich alloys: permalloys;

- amorphous cobalt rich alloys;

- non-compensated antiferromagnets: ferrites.

We will consider only the first two groups of materials because they are widely used in lownoise sensors.

6.1. COMMON REQUIREMENTS FOR THE MAGNETIC MATERIALS

A key parameter for the magnetic material to be used in the fluxgate core is the Barkhausen noise, which ultimately determines the limits for the sensitivity and the accuracy of the sensor. In modern magnetometers the analogue DC output is often directly converted by means of AD converters which today are available with up to at least 20 bits reliability (e.g. Sigma-Delta converters, CS 5321/22 from Crystal Semiconductor or AD 7710 from Analogue Devices). In the Earth's field (60 000 nT) the 20 bit resolution is $120\,000/2^{20} = 0.11$ nT. The corresponding acceptable sensor rms noise level in the relevant frequency band should be far below this value. $0.11/6 = 0.018$ nT is arbitrarily considered satisfactory. In the frequency range 0.1 to 10 Hz this corresponds to a spectral noise power density of 8.4 pT/$\sqrt{\text{Hz}}$. In extra-terrestrial missions with lower fields even lower noise levels may be required.

In the valuation of potential core materials various other properties may also be taken into consideration. The material properties will influence the performance of the sensor for which the following characteristics should be considered:

1. Sensor noise, i.e. essentially the Barkhausen noise.

2. Zero-field offset and its stability versus temperature.

3. Long-term stability of the offset.

4. Material uniformity and homogeneity, which are crucial for suppressing the odd harmonics in the sensor signal [Petersen et al. 1992].

5. Mechanical robustness and insensitivity to acoustic vibrations.

6. Insensitivity to transverse fields, i.e. to fields perpendicular to the axis of measurement.

7. Excitation power consumption.

All of these issues strongly depend on the choice and handling of the core material. Within the last 10–15 years a set of requirements has been defined which for magneto-soft materials will guarantee the best parameters for the sensors, (see volume I, 3.7). The set of requirements is:

$$\begin{array}{llll} \lambda_S \to 0; & K_1 \to 0; & \Theta \to 70\text{-}400°C; & B_S \to 0.2 - 0.3 \text{ T}; \\ B_R/B_S \to 0.5; & H_C \to & \mu_{\text{init.}} \to \max; & P \to \end{array} \quad (6.1)$$

where λ_S is the saturation magnetostriction coefficient; K_1 is the crystal anisotropy constant, Θ is the Curie temperature, B_S is the saturation induction, B_R is the remanent induction, H_C is the coercivity, $\mu_{\text{init.}}$ is the initial permeability ($\mu_{\text{init.}} \sim 1/H_C$) and P is the power losses on remagnetization.

The set of requirements (6.1) is valid both for crystalline and for amorphous alloys except for those amorphous alloys which are given a special stress annealing treatment, see (Sect. 6.2.1). However, for amorphous alloys too all requirements (6.1) may be satisfied simultaneously by the proper choice of amorphous alloy composition and treatment. The use of amorphous alloys in fluxgate sensors seems very promising.

Concrete ways to realize the set of requirements (6.1) are considered in the following sections. Let us only note here that a sensor noise level depends also on the way of remagnetizing the ferromagnetic core. This will be shown later (Sect. 6.2.1) for specific examples and it is an additional factor of strategy in the attempts of reducing the sensor noise level.

6.2 Crystalline Materials

Material Selection for Crystalline Alloys

A basic requirement for the soft magnetic materials is a symmetrical hysteresis curve in a wide temperature range. This can be assured by using crystalline soft magnetic materials with a homogeneous chemical composition and with a small impurity concentration for minimizing anisotropies. The desired physical parameters include a magnetic squareness ratio in the range $0.5\ldots0.8$ and a high electrical resistivity > 70 $\mu\Omega$ cm. The saturation magnetostriction coefficient λ_S should be within the range $+1$ to -2×10^{-6}.

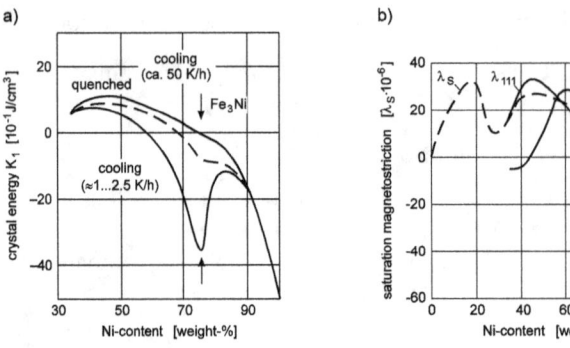

Figure 6.1: Composition dependence in the Fe-Ni system: (a) crystal anisotropy (b) magnetostriction.

In the Fe-Ni system the magnetic crystal energy K_1, (Fig. 6.1a) and the magnetostriction λ_S (Fig. 6.1b) have low values at 73% Ni and 81% Ni, respectively. By additions of Mo the zero passage of the crystal energy K_1 may be shifted towards 80% Ni and above it [Muller, 1978]. So in order to get high permeability and good magnetic noise behavior a 13Fe-81Ni6Mo-alloy (in Russia: 82NM) is used [Gordon, 1968; Acuña, 1974 and 1978; Afanassiev, 1979 and 1981; Fornaçon, 1993], where K_1 and λ_S are very small. Important is, that the values of λ_S are even weakly negative [Afanassiev, 1974; Afanassiev and Gorobey, 1981]. Besides there is a vanadium permalloy 12Fe-83Ni-5V (83NF in Russia) which in ringcores has lower noise level than the alloy (82NM). The adjustment of K_1 and λ_S in these alloys may be optimized by a special heat treatment in vacuum.

Ring Core Design and Manufacturing

The permalloy alloys are sequentially hot and cold rolled to 12.7 μm [Gordon et al., 1968; Infinetics] or 20 μm [Afanassiev and Gorobey, 1981; Fornaçon et al., 1993] in thickness with a deformation grade of nearly 98%, and then cut to 1.58 mm [Gordon et al., 1968; Infinetics], 1.5 mm [Afanassiev and Gorobey, 1981] or 2 mm [Fornaçon et al. 1993] in width. After etching of the cutting zone the tape is wound (normally 6–12 windings) and fixed on a bobbin. One end of the tape may be fixed on the bobbin using point electrowelding. Then the tape is wound under stress, and the end of the outer winding is electrowelded to the previous one. The mechanical precisely produced bobbin consists of a non-ferromagnetic high temperature stable alloy (Gordon et al. 1968 and INFINETICS Inc.:[1] Inconel X750, Afanassiev and Gorobey: X23, Fornaçon 1993: NiMo-30 and NiCrTiAl75.20) or a ceramic based material having the same linear thermal coefficient as the soft magnetic material. The electromagnetical insulation between the magnetic layers is made of an electrophoretical smooth MgO insulation sheet.

In a series of experiments at MPE (Max-Planck-Institut für Extraterrestrische Physik, Berlin-Adlershof)a number of soft magnetic materials were manufactured by the same technology. The typical parameters for the ringcores are: permalloy alloy 20 μm $\times$ 2 mm, 6 windings, diameter 13 mm, number of excitation windings 180, and excitation current amplitude 1 A peak-peak at 6.5 kHz. They were tested under the same conditions in a conventional laboratory μ-metal shield for determining the short-time noise up to a few minutes, and in a special zero point measurement facility, developed together with the Geomagnetic Observatory in Niemegk, for long-term stability studies. The experimental results are tabulated in table 1.

---

[1] INIFINETICS Inc., manufacturer of standard toroidal cores for ringcore fluxgate magnetometers.

As seen in the table a correlation exists between the saturation magnetostriction λ_S and the measured noise level of the ring core elements in the laboratory μ-metal shield in the frequency range from o.001 to 10 Hz, and the noise density at 1 Hz. The results show that the material with small negative saturation magnetostriction λ_S has the smallest noise level and the best long term stability Fig.6.2.

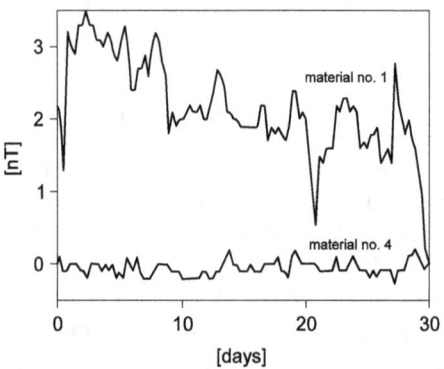

Figure 6.2: Long-term stability for alloys no. 1 and no. 4 from table 1

The soft magnetic material no. 4 has been used in the fluxgate magnetometers for the Mars 94/96 program MAREMF and MAGIBAL, for the main satellite Interball Tailprobe and the Aurora-sonde Relict-2, and for the Equator-S satellite.

It has to be remarked, that every ringcore element is an individual one. The smallest changes of the technology during the manufacturing can change the long term stability, the dependence on temperature and the noise behavior. Therefore it is necessary to choose the ringcore elements for space missions among a lot of produced species, and measure the main data from time to time. During the preparation for flight the sensors are stressed continuously with excitation fields and in a large temperature range between $-70°C$ and $+70°C$, with a temperature rate of about $1°C/min$).

The long-term stability under earth field conditions for most of the ringcore elements with material no. 4 is better than 1 nT and it is independent of the ringcore diameter (13 mm and 17 mm).

A Sensor Based on Thin Films – Design and Examination Results

A sensor of original construction, on the basis of a core consisting of a package of thin permalloy films, was developed by a staff at L. V. Kirenski Physics Institute (Krasnoyarsk, Russia) [Fzolov et al., 1987; Afanassiev, 1990]. The sensor diagram is shown in Fig. 6.3. The package represents a "puff pie" of six films (15×15 mm, thickness 100 nm each) made by vacuum spraying from the material 81Ni-19Fe, having $\lambda_S \sim 0$. The package is enveloped by two inter-perpendicular coils one of which, having the axis parallel to the direction of the field B_{0x} to be measured, is a driving coil. Another one, which is perpendicular to the measured field, but is parallel to the bias field from the two magnets, is a pick-up coil.

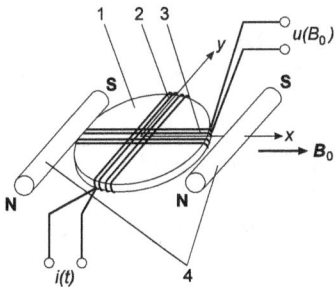

Figure 6.3: The thin film sensor described in the text.

The field from the magnets magnetizes the films in the direction of the y-axis up to full magnetic saturation (mono-domain state). Now, if $B_{0x} = 0$ then the superposition of the alternating driving field leads to a "swinging" symmetric with respect to the y-axis of the film magnetization vector $\mathbf{J}_S$. This causes the induction of an EMF $e_{2f}(t)$ of double frequency in the pick-up coil, see Fig. 6.4a.

If $B_{0x} \neq 0$ the "swings" of vector $\mathbf{J}_S$ will be asymmetric, and together with the EMF of double frequency $e_{2f}(t)$, an EMF of basic frequency $e_f(t)$ occurs in the pick-up coil, too – see Fig. 6.4b. This EMF amplitude gives information about the measured field B_{0x}. A magnetometer prototype on the basis of this sensor was examined in 1990 at D. I. Mendeleyev Institute for Metrology (St. Petersburg, Russia) (with the participation of Y. V. Afanassiev). It was found that the transformation ratio of the magnetometer was 8.3 mV/nT with a total consumption of power less than 0.1 W (sensor driving frequency $f = 1.1$ MHz). At these conditions, the spectral density of the magnetometer noise was as follows:

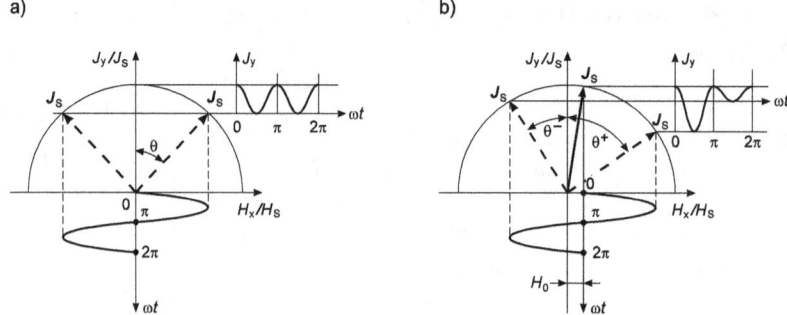

Figure 6.4: Thin film sensor. Principle of operation: (a) external field $B_{0x} = 0$. (b) external field $B_{0x} \neq 0$.

>   0.5 Hz: 2.7 pT/$\sqrt{\text{Hz}}$;   100 Hz: 1.0 pT/$\sqrt{\text{Hz}}$;
>   1.0 Hz: 2.2 pT/$\sqrt{\text{Hz}}$;   1000 Hz: 0.8 pT/$\sqrt{\text{Hz}}$;
>   10 Hz: 1.3 pT/$\sqrt{\text{Hz}}$.

This is not a bad result for a noise level, and it shows evidently that together with a correct selection of the magnetic material (with respect to the set of requirements (6.1)), it is still possible to reduce the sensor noise level, by using a proper core driving method. In the present case the way is used where the core remagnetization is provided by a coherent rotation of the magnetization vector. Here the core (film package) magnetization curve is free of hysteresis in the x-axis direction. A drawback of the sensor is the high noise level near zero frequency because of the direct penetration of the driving field into the pick-up circuit. However, this effect will not be so essential, if the task is searching for very small alternating magnetic fields in the frequency range 1–10000 Hz.

6.2.1 Amorphous Materials (Metallic Glasses)

Ferromagnetic metallic glasses are produced in the laboratory as thin ($\sim$20–30 μm) sheets or ribbons by rapid quenching ($\sim 10^{6}\,^{\circ}C/s$) of melted alloys with approximate eutectic compositions like $(\text{FeCoNi})_{75}(\text{BSiP})_{25}$. Here the transition metals (FeCoNi) are mainly responsible for the magnetic properties, while the metalloids (BSiP) constitute a so-called glassformer maintaining the liquid structure during the rapid cooling process.

It has been shown [Shirae, 1984] that a noise density as low as 1.1 pT/$\sqrt{\text{Hz}}$ at 1 Hz ($\sim$2.4 pT rms 0.1–10 Hz assuming $1/f$ noise power spectral dependence) is obtainable with amorphous $\text{Co}_{62.4}\text{Fe}_{2.8}\text{Cr}_{7.0}\text{Si}_{13.9}\text{B}_{13.9}$ ribbons in a

Vacquier type sensor. This material, however, has a Curie temperature of only 50°C, which is non-acceptable for most magnetometers. Another composition $Co_{66}Fe_4Si_{12}B_{18}$ investigated as ring cores by Narod et al. [1985] showed a noise level of 10 pT/$\sqrt{Hz}$ at 1 Hz ($\sim 21\ pTrms 0.1--10\ Hz$).

Studies of the noise dependence on the magnetostriction were made by Nielsen et al., [1990] in a series of alloys $(Co_{1-x}Fe_x)_{70}Si_{12}B_{18}$ including the alloy studied by Narod et al. [1985]. A pronounced minimum for the noise (17 pT rms 0.06–10 Hz ~ 7.9 pT/$\sqrt{Hz}$) was found for $Co_{66.5}Fe_{3.5}Si_{12}B_{18}$ ($x = 0.05$) in which composition the magnetostriction is zero after an appropriate stress annealing process. Furthermore it was found that the commercially available VITROVAC 6025 ($(CoFe)_{70}(MoSiB)_{30}$) behaves essentially like the $Co_{66.5}Fe_{3.5}Si_{12}B_{18}$ alloy.

Glass Compositions

As mentioned above a low (zero) magnetostriction is desirable for the fluxgate core material. In the (FeCoNi)(BSiP) metallic glass system all compositions except the most Co-rich have positive magnetostriction. However, if the content of Fe+Ni is below roughly 5 percent of the total amount of transition metals, the magnetostriction is negative. This means that in the triangular FeCoNi composition diagram near the Co-rich corner there is a boundary line between positive and negative magnetostriction of which compositions the magnetostriction is zero.

In order to optimize its fluxgate properties the as-quenched ribbon may need a heat treatment which affects the magnitude of the magnetostriction. Figure 6.5 shows for a series of ribbons of $(Fe_xCo_{1-x})_{70}Si_{12}B_{18}$ the composition dependence of the magnetostriction coefficient λ_S in the as-quenched state, and after an appropriate heat treatment, see below. From Fig. 6.5 it is concluded that taking the heat treatment into consideration the most suitable fluxgate composition is the one with $x = 0.5$, i.e. $Co_{66.5}Fe_{3.5}Si_{12}B_{18}$. The figure also shows the behavior of the commercial alloy VITROVAC 6025. It is seen that the heat treatment used in Fig. 6.5 causes an increase of λ_S by $0.2 \cdot 10^{-6}$ which leaves this VITROVAC 6025 with a positive magnetostriction of $0.07 \cdot 10^{-6}$. The VITROVAC alloys, however, may differ from batch to batch because the composition of each batch may have been adjusted to fulfill some specific annealing conditions required by the costumers. Therefore for VITROVAC it is necessary to know the annealing temperature dependence of the magnetostriction in order to choose annealing conditions which leaves the material with zero magnetostriction. This dependence is shown in Fig. 6.6 which reveals a zero crossing of λ_S for the annealing temperature 450°C.

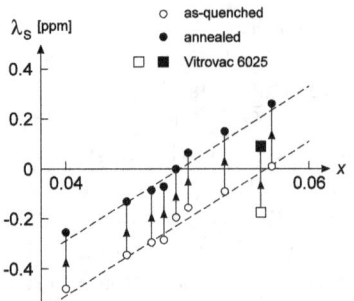

Figure 6.5: Composition dependence of the magnetostriction coefficient λ_S for as-quenched ○ and for annealed ● ribbons of $(Fe_xCo_{1-x})_{70}Si_{12}B_{18}$. Annealing conditions: 1 h pre-annealing at $360°C$ followed by 1 h stress-annealing at $340°C$ with applied stress = 400 MPa.

From Figs. 6.5 and 6.6 it is concluded that metallic glasses of different compositions require different annealing conditions in order to obtain optimum fluxgate properties. Although Fig. 6.6 represents the behavior for only one specific composition, we believe that other Co-rich metallic glasses with near-zero magnetostriction behave in a similar manner.

The $Co_{66.5}Fe_{3.5}Si_{12}B_{18}$ alloy has very good glassforming properties, and it is easily fabricated by the single roller quenching method in the laboratory as long, 1 mm wide, ribbons. Its Curie temperature is $260°C$.

VITROVAC 6025 is available as 1 mm wide, 0.025 mm thick, long ribbons from Vacuumschmelze GmbH, Hanau, Germany. The Curie temperature of this material is $250°C$.

Glassy Materials Processing

In order to improve the fluxgate properties the as-quenched ribbons may need an annealing treatment which stabilizes the material structure at the same time as the Barkhausen noise is reduced. One possible treatment may be performed in two steps as follows: i) pre-annealing (stress relief) and ii) stress-annealing. The effect of annealing on the magnetic properties of metallic glasses is described in a review paper by Nielsen [1985].

The structural stabilization (stress relief) is obtained by a pre-annealing of about 1/2 to 1 h at a temperature in the range $320\,to\,450°C$. The time and temperature are not very critical parameters, but the lower annealing temperature the longer annealing time is required.

The Barkhausen noise is reduced by producing a magnetic anisotropy

6.2. CRYSTALLINE MATERIALS

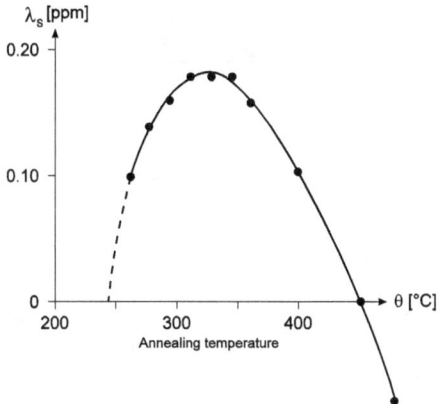

Figure 6.6: Annealing temperature dependence of the magnetostriction coefficient λ_S for VITROVAC 6025 measured at room temperature. Below 345°C 1 h pre-annealing followed by 1 h stress-annealing at the same temperature. Above 345°C 1 h pre-annealing at the temperature shown by • followed by 1 h stress annealing at 350°C. Applied stresses 150 MPa.

of which the hard axis is parallel to the ribbon axis. The anisotropy is a result of a structural property which is generated by the unelastic creep developing during a stress-annealing process. The creep is completed after annealing about 1/2 to 1 h in the temperature range 300°C to 400°C under applied tensile stress of about 200 MPa. As for the pre-annealing the stress-annealing parameters are not very critical for inducing the anisotropy but the most efficient annealing temperature is around 340°C. Figure 6.7 shows the magnetization curve for a VITROVAC 6025 ribbon treated in this way.

The magnetization curve in Fig. 6.7 is virtually hysteresis-free which means that the magnetization process is expected being due to coherent rotation of the spins within the domains without movements of the domain walls. An electron microscope study of the domain structure confirms this hypothesis.

Figure 6.8 shows the domain pattern in a stress annealed ribbon in zero field. If a field is applied along the ribbon axis, however, the pattern does not change until it disappears at saturation. This means that the magnetization process takes place without wall movements, thereby reducing the Barkhausen noise. Furthermore this domain structure makes it possible to produce fluxgates which are almost *completely non-sensitive to transverse fields*.

As indicated above the values of the annealing parameters are not very

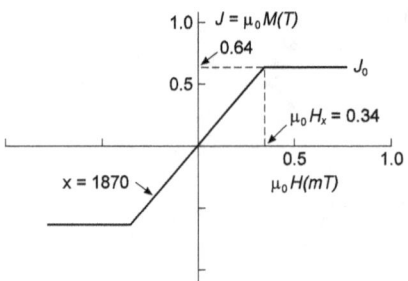

Figure 6.7: Magnetization curve for a metallic glass ribbon of VITROVAC 6025 which has been annealed as described in the text: pre-annealing 1 h at 420°C followed by 1 h stress-annealing at 340°C with applied stress = 200 MPa.

critical for obtaining good fluxgate properties. It is worth noting, however, that the anisotropy field (H_a in Fig. 6.7) achieves a maximum value for stress-annealing temperatures around 340°C. Furthermore the anisotropy field saturates within the first hour of stress-annealing, and its value is proportional to the tensile stress during the annealing process.

The purpose of the pre-annealing is first to suppress, during the subsequent stress-annealing, a plastic creep which may reduce the magnitude of the induced anisotropy. Secondly it is possible by the pre-annealing to adjust the magnetostriction to zero (Fig. 6.6). Furthermore we expect the pre-annealing to improve the long-term stability.

Finally we conclude that a suitable annealing procedure for a VITROVAC 6025 ribbon consists of 1/2 h pre-annealing at 450°C (stress relief and promotion of zero magnetostriction) followed by 1/2 h stress-annealing at 340°C with applied stress of 200 MPa (induction of a uniaxial hard ribbon axis anisotropy).

Metallic Glasses as Ringcores

The metallic glass ribbons produced and processed as described above need no special precautions during storage and mounting in the excitation current bobbin. Due to their nonmagnetostriction and their high ductility they are almost unaffected by mechanichal strains and other "maltreatments". This means that after mounting in the bobbin the core does not need a final heat treatment as is the case for permalloy alloys, especially in ringcore sensors.

By using metallic glasses as ring cores we may benefit from their very special mechanical and magnetic properties. Considering the ringcore sensor, it is important to maintain the rotational symmetry as perfect as possible.

6.2. CRYSTALLINE MATERIALS

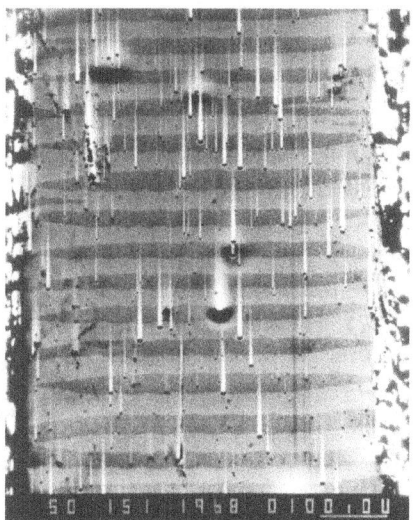

Figure 6.8: Domain structure of a stress annealed metallic glass ribbon. The teeth in the comb pattern extend from one edge to the other in the 1 mm wide ribbon.

Because of the "spring steel character" of the ribbon the straight ribbon may be mounted in the inner groove of a ring shaped bobbin. The spring force keeps the wraps in a circular position except for the innermost wrap which ends as an arc of a circle.

The innermost wrap may be given a circular shape by means of an "open ring" shaped hold-down spring mounted inside the ribbon. However, in order to maintain the symmetry it is desirable and important that the radial force density is as uniform as possible, which means that the hold-down spring must be tailored to this condition.

The uniform radial force density is obtained by choosing a suitable ring cross section area. A strain-stress analysis shows that the angular variation of the ring thickness t should follow the following formula [Nielsen et al., 1996]:

$$t^3 = \frac{3pd^4}{2Ewd} sin^2 \frac{\theta}{2} \qquad (6.2)$$

where p is the uniform force density (N/m), E is Youngs modulus, w is the width, d is the ring diameter and Δ is the numerical change of the diameter when from the free to the mounted condition. The hold-down rings may be produced of stainless steel or an Inconel alloy, and they are easily produced by punching in a computer controlled workshop machine.

A typical fluxgate ring consists of a bobbin made of the machinable glass MACORR supplied with a 1 mm wide annealed VITROVAC 6025 which is kept in position by an Inconel 625 hold-down ring. Bobbin width = 2.9 mm, outer diameter = 18.5 mm, inner diameter = 15.0 mm, groove width = 1.1 mm, groove depth = 1.0 mm. Number of wraps = 12 corresponding to a core cross section area = 0.3 mm^2. The hold-down ring which is 1.0 mm wide with a maximum thickness of 0.4 mm has an outer diameter = 17.5 mm in the nonstrained (open) condition.

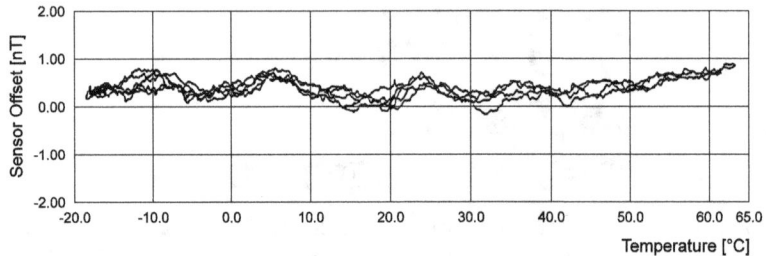

Figure 6.9: Thermal variation of offset for a metallic glass sensor, during two full temperature cycles from -20 to $+20°C$.

Figure 6.9 shows the behavior for a fluxgate sensor which is mounted with a ring produced as described above. The sensor which is placed in a magnetic shield is exposed to a repeated thermal cycling between -20 and $+60°C$. As seen from the figure the offset keeps below 1 nT over the whole temperature range, deviating at most 0.5 nT from its mean value 0.4 nT. The deviation is reproducible within 0.2 nT in the whole temperature range. At constant temperature the sensor noise power density is 6 pT/$\sqrt{Hz}$ at 1 Hz. We emphasize that this behavior is typical for a sensor mounted in the magnetometer without any adjustments to individual sensors after the mounting. Sensors of this type are used in the Danish geomagnetic research microsatellite [Nielsen et al. 1995].

Amorphous Alloys Having Superhigh Magnetic Permeability

The high-cobalt alloy $Co_{64}Fe_3Si_{16}B_{12}Cr_5$ was examined experimentally [Afanassiev and Sherbackova, 1995]. The permeability curve (1) of this alloy is compared with the permeability curve of a 6%-Mo permalloy (2) in Fig. 6.10. It is seen

6.2. CRYSTALLINE MATERIALS

that the amorphous alloy is characterized by an anomalous high permeability at very low magnetic fields. Other characteristics of the amorphous alloy are:

$$\lambda_S \leq 10^{-6}; \quad B_R/B_S = 0.54;$$
$$\Theta = 92°C; \quad H_C = 0.2 \text{ A/m};$$
$$B_S = 0.24 \text{ T}; \quad \mu_{init.} = (80 - 100)10^3.$$

These data show that the alloy is very good for the satisfaction of the requirement set (6.1 in Sect.6.1).

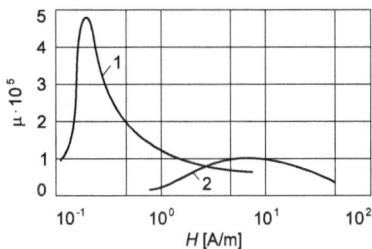

Figure 6.10: Permeability vs. excitation field for 1) amorphous $Co_{64}Fe_3Si_{16}B_{12}Cr_5$ and 2) 6% Mo permalloy.

Curves of spectral noise density are shown in Fig. 6.11a for amorphous (1) and permalloy (2) ring cores respectively. The spectral noise density level of the amorphous cores at frequencies higher than 10 Hz was found 2–3 times lower than that of the permalloy cores, at driving field amplitude 2.5 kA/m. The dependence of the spectral noise density at frequency 10 Hz

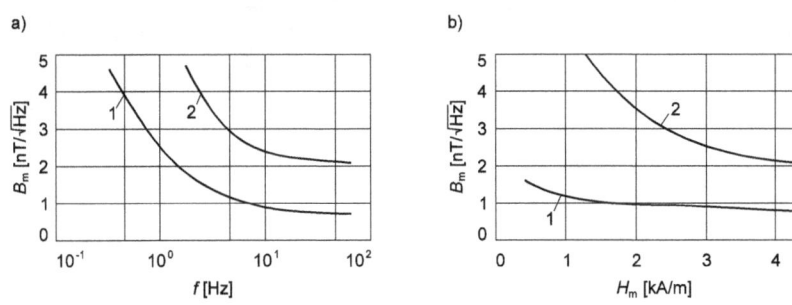

Figure 6.11: Spectral noise density for 1) $Co_{64}Fe_3Si_{16}B_{12}Cr_5$ and 2) 6% Mo permalloy. (a) noise with driving field amplitude = 2.5 kA/m (b) noise density at 1 Hz vs. driving field amplitude.

on the variation of the driving field amplitude is shown in Fig. 6.11b. It is seen that the amorphous cores (curve 1) may operate at relatively lower driving field amplitudes than the permalloy cores (curve 2).

The cores were annealed at the temperature $350°C$ in 30 minutes. It is interesting to note for this alloy that approximately the same results for noise levels were found for non-annealed ring cores. In our mind, this alloy is very promising for use in fluxgate sensors.

References chapter 6

1. Acuña, M.H., Fluxgate magnetometers for outer planets exploration, IEEE Trans. on Magn., Vol. 10, No. 3, 519-523, 1974.

2. Acuña, M.H., AC.S. Scearce, J.B. Seek, J. Scheifele, The Magsat vector magnetometer a precision fluxgate magnetometer for the measurement of the geomagnetic field, NASA Technical Memorandum 79656, 1978.

3. Afanassiev, Yu.V., Perspectives of further noise level reducing at ferromodulation transducers, Measurement technics, N1, 1990

4. Afanassiev, Yu.V. V.N. Gorobey, Noises of fluxgates and magnetic amplifiers, Measurements, Control, Automatisation, N4, 37-53, 1981.

5. Afanassiev, Yu.V., T.I. Sherbackova, New research of fluxgate sensors noise: amorphous alloys cores are bester than permalloy ones. An articles collection Barkhausen effect and similar physical phenomena" Iszevsk, 159-164, 1995.

6. Afanassiev, Yu.V., N.V. Studentenv, V.N. Horov, E.N. Chenhurina, A.P. Shelkin, Means of measurements of magnetic field parameters, L.: Energia, 1979.

7. Fornaçon, K.H., M. Müller, I. Fischer, A magnetic sensor on the basis of a Fe-80Ni-6Mo alloy for magnetic field measurements in space, Soft Magnetic Materials Conference 10, Venedig, P6-39, 1993.

8. Fzolov, G.I., V.V. Poliakov, A.G. Vladimirov, Magnetomeasuring transformers based on thin films, Interorganization collection of scientific words of AN USSR.
KKzuenoyarsk: Physics Institute im. L VKirenetogo, 37-50, 1987.

9. Gordon, D.I., R.H. Lundsten, R.A. Chiarodo, H.H. Helms, A fluxgate sensor of high stability for low field magnetometry, IEEE Trans. on Magn., Vol. 4, No. 3, 397-401, 1968.

10. INFINETICS, INC, manufacturer of standard toroidal cores for ring-core fluxgate magnetometers.

11. Müller, M., R. Steinelt, Magnetisch hochpermeable Legierungen, Stahlberatung, Vol. 5, 3-8, 1978.

12. Narod, B.B., J.R. Bennest, J.C. Strom-Olsen, F. Nezil and R.A. Dunlap, An evaluation of the noise performance of Fe, Co, Si, and B amorphous alloys in ring-core fluxgate magnetometers, Can. J. Phys. 63, 1468-1472, 1985.

13. Nielsen, O.V., Effects of longitudinal and torsional stress annealing on the magnetic anisotropy in amorphous ribbon materiale, IEEE Trans. on Magn. MAG21, 2008-2013, 1985.

14. Nielsen, O.V., J.R. Petersen, B. Hernando, J. Gutierrez and F. Primdahl, Metallic glasses for fluxgate applications, Anales de Fisica B86, 271-276, 1990.

15. Nielsen, O.V., J.R. Petersen, F. Primdahl, P. Brauer, B. Hernando, A. Fernandez, J.M.G. Merayo amd P. Ripka, Development, construction and analysis of the 'Oersted' fluxgate magnetometer, Meas. Sci. Technol. 6, 1099-1115, 1995.

16. Nielsen, O.V., P. Brauer, F. Primdahl, J.L. Jorgensen, C. Boe and S Bauereisen, A highprecision fluxgate sensor for space applications: layout and choice of materiale, to be published.

17. Petersen, J.R., F. Primdahl, B. Hernando, A. Fernandez and O.V. Nielsen, The ring core fluxgate sensor null feed-through signal, Meas. Sci. Technol. 3, 1149-1154, 1992.

18. Shirae, K., Noise in amorphous magnetic materiale, IEEE Trans. Magn. MAG-20, 1299-1301, 1984.

Chapter 7

Magnetic Field Vector Feedback

F.Primdahl,

DanishSpaceResearchInstitute,
Lyngby,Copenhagen

7.1 Introduction

The B-field generated (open loop) output signal from a fluxgate sensor depends on the (demagnetized) external magnetic field inside the sensor core, and it is only strictly proportional to B for fields much smaller than the saturation field of the magnetic core material. For larger fields the nonlinearity is noticeable, and the open loop sensor output in fact reproduces the curvature of the core material hysteresis loop (Acuña and Pellerin, 1969; Pellerin, Acuña, 1970; Acuña and Ness, 1975).

After stress annealing the amorphous metal magnetic materials show a piece-wise linear magnetization curve (see Fig. 2 of Nielsen et al., 1995), and Fig.7.1 shows the open loop response of a high sensitivity ring core sensor of this material. However, sensor linearity alone is not sufficient, because the open loop output also depends on the susceptibility of the core material, and this parameter is highly temperature sensitive leading to an unacceptably large temperature coefficient of the magnetometer output.

Applying a feedback magnetic field along the sensor axis, then the sensor is used as a null-detecting device. The advantage of using this balancing

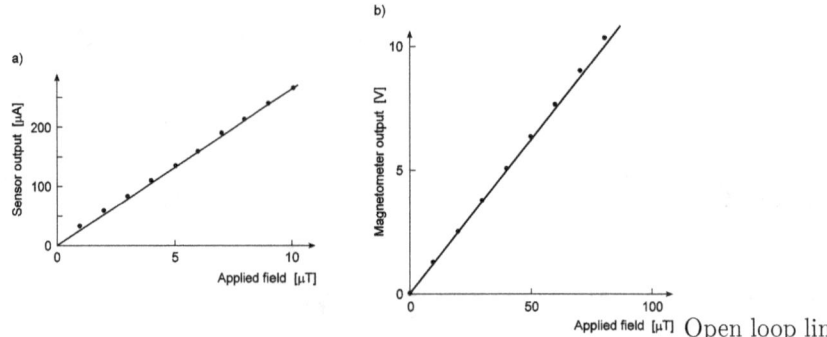

Open loop linearity of the stress annealed amorphous metal ring core sensor: a)Peak-to-peak short circuited sensor AC output current against input field, and b)open loop DC output voltage of the magnetometer.

technique is that the null of the sensor is far more stable over time and over a temperature range than is the sensor output at some non-zero axial field. The linearity of the fluxgate magnetometer in a high gain feedback configuration is observed to be better than $4-6$ relative to the FSR (full scale range), probably limited by the test field reproducibility rather than by the sensor feedback loop (Primdahl, Lühr and Lauridsen, 1992).

For practical reasons, and in order to save weight, volume and power, the sensor secondary coil is generally used to supply the feedback nulling field, and at the same time, for detecting the sensor output signal. However, the use of a separate feedback coil may in certain cases be advantageous.

If the sensor core changes position inside the feedback coil, then any coil field inhomogeneity will cause the average field over the core volume to change, and a change in the magnetometer sensitivity will result. The feedback coil should then create a highly homogeneous nulling field, which is basically contradictory to the need for a close coupling of the pickup coil to the core magnetic flux. Furthermore, a large inhomogeneous feedback field will change the output signal phase (Felch and Potter, 1953), and for different external fields the average field weighted over the core volume may not be strictly proportional to the feedback coil current.

Unless special precautions are taken when using a single coil, then the output impedance of the feedback current source will load the coil and attenuate the sensor output signal. Also, residual even harmonic signals in the feedback current will couple directly to the input of the component amplifier, and if too large they may constitute a risk of introducing instability in the instrument.

In summary, the use of a special feedback coil separated from the fluxgate

7.1. INTRODUCTION

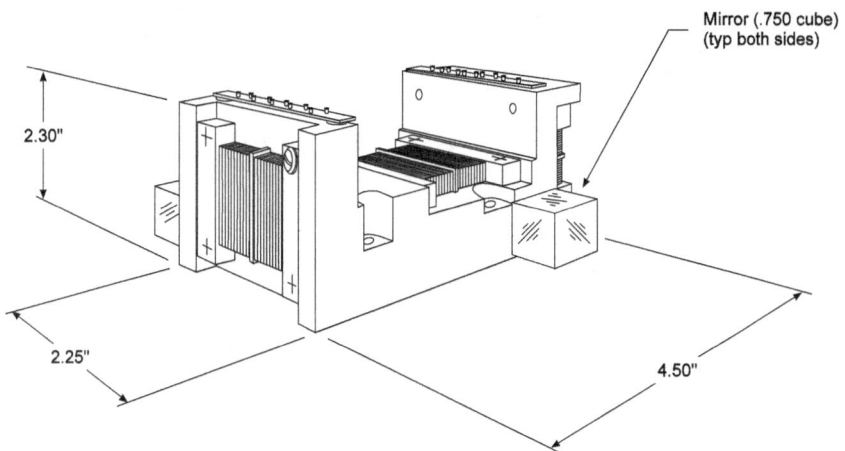

Figure 7.1: Tri-axial single axis compensated fluxgate sensor.

secondary pickup coil avoids a number of potential error sources which are present in the single coil feedback system. This at the cost of a mechanically more complex structure and a more power demanding and heavier sensor. The choice in each case depends on careful considerations of the real needs for instrument precision and stability that are necessary to meet the specific task requirements.

7.1.1 Triaxial Feedback Configuration

In a tri-axial fluxgate sensor configuration the three individual fluxgates are normally placed close together having the magnetic axes approximately at right angles to each other. Small angular deviations from orthogonality between the sensor axes are of little concern, as long as they are stable. Tri-axial sensor axes have deviations of the order of $0.1° to 0.5°$, which can easily be corrected for by using a sensor calibration matrix.

Figure 7.1 shows the Magsat magnetometer sensor (Acuña et al., 1978) as an example of a widely used sensor configuration. The sensors are placed symmetrically with respect to each other with one sensor at a larger distance from the two other sensors in order to minimize the DC-field interference between the sensors.

The external field disturbance from the feedback currents in the secondary coils is small because the coil dimensions are small. With 60 000 nT inside the coil, the external field will typically be less than 1 nT at a distance of about 20 coil diagonals (approximately 50 cm), because the field decreases

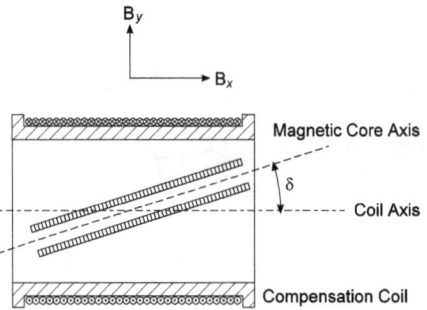

Figure 7.2: Single axis compensated sensor with the core axis deviating from the feedback coil axis

as the third power of the distance from the coil. This is important in a satellite instrument for geomagnetic field mapping where an absolute scalar magnetometer sensor may be placed close to the vector sensor for calibration purposes.

In a single axis fluxgate sensor the feedback compensation coil axis as well as the core magnetic axis will influence the magnetic axis of the sensor. Figure 7.2 schematically shows a single axis sensor, where the core axis tilts an angle δ with respect to the coil axis. The coil current compensates for the B_x field, but B_y also has a component $B_y \sin \delta$ along the core. In a high gain feedback configuration the magnetometer reacts by adding an extra field ΔB_x along the feedback coil axis, and the component of this extra field along the core axis is:

$$\Delta B_x \cos \delta = -B_y \sin \delta \ . \tag{7.1}$$

The resultant sensor output B_{out} is then:

$$B_{\text{out}} = B_x - B_y \tan \delta \ . \tag{7.2}$$

A field B_0 of constant magnitude and rotating in the x-y-plane will then give the following sensor output response:

$$B_{\text{out}} = B_0 \cos(\Omega t) - B_0 \sin(\Omega t) \tan \delta = \frac{B_0}{\cos \delta} \cos(\Omega t + \delta) \ . \tag{7.3}$$

This shows that the sensor magnetic axis is rotated the angle δ, and the sensitivity is changed by the factor $1/\cos \delta$. As long as δ is stable, then this is of no concern, but any mechanical shifts in the relative positions of the components will affect the stability of the sensor. Note that this result is valid for a sensor in an infinite gain feedback loop, for a sensor in open loop configuration the magnetic axis follows the core magnetic axis, and the sensitivity will be attenuated by the factor $\cos\delta$.

7.1.2 Vector Feedback Systems

The ring core of a fluxgate sensor responds symmetrically to field vectors in the plane of the ring, whereas it is insensitive to fields perpendicular to the ring plane because of the very large demagnetizing factor in this direction. Single axis compensated ring cores are exposed to the full external field perpendicular to the compensation coil axis, and the transverse field nonlinearly influences the on-axis sensitivity. Sensors of a linear geometry such as double-rod and "helicoil" sensors do not have this asymmetry because of the large demagnetization transverse to the sensor axis.

The Magsat magnetometer used single axis compensated ring core sensors, and the transverse field effect of the order of tens of nT in full Earth's field was first described, calibrated and fully modelled by Acuña (1981). In the Magsat study report Adams et al. (1976) recommended the use of a triple system of coils for simultaneously nulling the field at the sensors in all directions, and it was shown that in this system the sensitivity to sensor misalignment is greatly reduced. They recommended a system of three Hemholtz coils as the best choice after rejecting spherical coils for geometrical reasons. However, for Magsat this was considered to be too large for maintenance at a constant temperature onboard the satellite mission, and the chosen magnetometer sensor used three orthogonal single axis ring core sensors. This was, of course, before the discovery of the transverse effect in single axis compensated ring core sensors by Acuña (1981).

Afanassiev and Bushuev (1978) developed a box shaped triaxial sensor ($11 \times 12 \times 13$ cm) with a single tubular core sensitive to the magnetic field in three orthogonal directions. The outer faces supported three orthogonal sets of square Helmholtz feedback coils simultaneously vector nulling the field at the core. The single-core configuration ensured that measurements of the field components referred to a single point in space.

Primdahl and Jensen (1982) described the development of a compact spherical coil (CSC) for fluxgate vector feedback based on an earlier paper by Everett and Osemeikhian (1966), who used a set of two concentric spherical coils for generating bias fields for an observatory proton magnetometer. Everett and Osemeikhian solved the geometrical problem of access to the inside of the coils by building each coil in two half shells, which could be assembled around the magnetometer sensor. The CSC fluxgate by Primdahl and Jensen (1982) consists of three concentric spherical coils of slightly different diameters in order to accommodate the windings, and each coil was made up of 9 individual coaxial circular coils on the spherical surface (see Fig. 7.1).

In a report on candidate magnetometers for Earth's field studies LaLanne (1990) described a triaxial vector feedback coil with three orthogonal sets of spherical feedback coils, each set consisting of 6 circular multiturn windings.

7.1.3 Stability of the Vector Feedback System

Figure 7.3 shows a two-axis field nulling system indicated by the two orthogonal sets of coaxial circular coils. It is fairly simple to demonstrate the

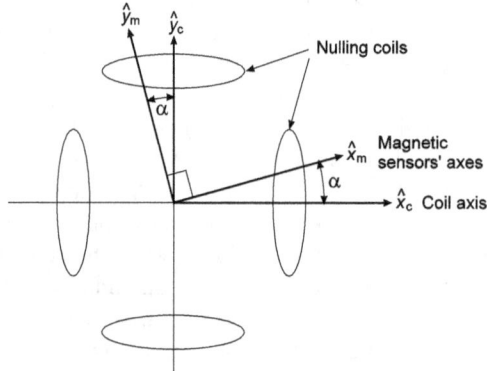

Figure 7.3: Two-dimensional vector feedback coil system.

independency of sensor orientation for this two-dimensional case. The coil system have axes x_c and y_c, and the fluxgate sensors have the axes x_m and y_m rotated a small angle α. For simplicity the coil and the fluxgate systems are assumed orthogonal. The coil fields are $B_{x,c}$ and $B_{y,c}$, and the external fields are B_x and B_y. The coil fields along the fluxgate magnetic axes $B_{x,m}$ and $B_{y,m}$ are nulled by the feedback loops of the magnetometer.

$$\left. \begin{array}{rcl} B_{x,m} &=& B_{x,c}\cos\alpha + B_{y,c}\sin\alpha - (B_x\cos\alpha + B_y\sin\alpha) = 0 \\ B_{y,m} &=& B_{y,c}\cos\alpha + B_{x,c}\sin\alpha - (B_y\cos\alpha + B_x\sin\alpha) = 0 \end{array} \right\} \quad (7.4)$$

Using $\cos\alpha \approx 1$ and $\sin\alpha \approx \alpha$, we seek solutions for $B_{x,c}$ and $B_{y,c}$ expressed by expressed by B_x, B_y and α:

$$\left. \begin{array}{rcl} B_{x,c} &=& \dfrac{1}{D}\left(B_x + \alpha B_y + \alpha^2 B_x - \alpha B_y\right) = B_x \\ B_{y,c} &=& \dfrac{1}{D}\left(-\alpha B_x + B_y + \alpha B_x + \alpha^2 B_y\right) = B_y \end{array} \right\} \quad (7.5)$$

7.1. INTRODUCTION

where $D = 1 + \alpha^2$ is the system determinant. The angle α thus drops out of the equations.

This result also holds for the three-dimensional case, and for the case where the individual fluxgate sensors are not orthogonal to each other, and where the coil system axes are skewed, and it holds even for large deviation angles.

The stability of a vector feedback system then depends only on the coil system stability and not on the stability of the fluxgate sensor axes.

References chapter 7

1. Acuña, M.H. and C.J. Pellerin, A Miniature Two-Axis Magnetometer, IEEE Trans. Geosci. Electronics, Vol. GE-7, 252-260, 1969.

2. Acuña, M.H. and N.F. Ness, The Pioneer XI High Field Fluxgate magnetometer, Space Science Instrumentation, Vol. 1, 177-188, 1975.

3. Acuña, M.H., C.S. Scearce, J.B. Seek and J. Scheifele, The MAGSAT Vector Magnetometer - A Precision Fluxgate Magnetometer for the Measurement of the Geomagnetic Field, NASA Technical Memorandum 79656, 1-18, NASA GSFC,Oct. 1978.

4. Acuña, M.H., MAGSAT - Vector Magnetometer Absolute Sensor Alignment Determination, NASA Technical Memorandum 79648, NASA Goddard Space Flight Center, Greenbelt MD 20771, U.S.A., 1981.

5. Adams, D.F., U.G. Hartman, L.L. Lazarow, J.O. Maloy and G.W. Mohler, Vector Magnetometer Design Study, etc., Final Report NASA CR-144825, 1976.

6. Felch, E.P. and J.L. Potter, Preliminary Development of a Magnetor Current Standard, Trans. Am. Inst. Elect. Engrs., Vol 72 part 1, 525-531, 1953.

7. LaLanne, F.-X., R. Verhille, G. Guillemin and I. Hrvoic, GEOMAG, New Development in Magnetometry for Earth's Science Studies, Institut de Physique du Globe de Paris, 4, Place Jussieu F-75252, cedex 05, France, 1990

8. Pellerin, C.J. and M.H. Acu!a, A Miniature Two-Axis Magnetometer, NASA Technical Note, NASA TN D-5325, NASA Washington, D.C., U.S.A., Jan. 1970.

9. Primdahl, F., H. Lühr and E. Kring Lauridsen, The Effect of Large Uncompensated Transverse Fields on the Fluxgate Magnetic Sensor Output, DRI Report 1-92, 1-21, Danish Space Research Institute, 1992.

Chapter 8

Digital Magnetometer

F.Primdahl[1], *H.Lühr*[2]

[1]*DanishSpaceResearchInstitute, Lyngby,Copenhagen*,
[2]*GeoForschungszentrumPotsdam, (GFZ), Berlin*

8.1 Digital Detection and Feedback Fluxgate Magnetometry

8.1.1 Introduction

Modern data acquisition and telemetry systems demand digital communication with the electronics systems supporting a front-end measuring transducer.

In space magnetometry this is traditionally done by adding an analogue-to-digital converter (A/D) to the analogue feedback loop output, often combined with ranging or stepping capabilities to make up for a limited dynamic range of the previously available A/D-circuits. Flexible adaptation to the telemetry systems is then handled by a data processing unit with storage, time tagging capability, and possibly also data compression routines (Balogh et al., 1993; Nielsen et al., 1995).

In a modern vector magnetometer the analogue feedback loops of the three components use a very modest fraction of the total system mass and power. Still, the overall performance of a full Earth's field instrument depends heavily on the analogue circuits, because the accuracy and stability of the analogue feedback loops, rather than those of the sensors, will limit the dynamic range and stability. This provides an incentive to replace the analogue signal conditioning circuitry by precise mathematical algorithms. A high-performance digital signal processor (DSP) is padded between a fast sampling input A/D digitizing the broad band sensor signal, and an output digital-to-analogue converter (D/A) providing the field nulling current. All analogue signal processing is then replaced by mathematical routines limited in performance only by the number of digits and by the clock speed of the DSP, and both are certain to increase in the future.

Forerunners for a digital feedback loop magnetometer are the offset incrementing systems for extending the full scale range of an analogue loop by step wise suppression of the zero of the instrument. One such system (McPherron et al., 1975) used an up/down counter to generate the offsets controlled by the output levels of the basic low-range magnetometer.

The technique of digitizing the signal as close to the source as possible is widely used in other transducer applications. The reasons for its late introduction into fluxgate magnetometry has a lot to do with traditions in space instrumentation, where designs are derived from a long heritage of successful experiments, and tend to be highly conservative. The introduction of digital feedback is a major departure from the customary way of doing things; but it takes advantage of the rapid development in digital techniques, and no doubt it represents the future in high precision space magnetometry.

8.1.2 The B-Field Related Sensor Output Signal

The broad band fluxgate sensor output consists of a superposition of signals at the fundamental frequency and at the higher harmonics of the core excitation current. A simple description is that the even harmonics contain information about the external magnetic field along the sensor axis and the odd harmonics constitute the feed-through signal coupled to the secondary coil from the excitation coil and from the magnetized core (Petersen et al., 1992). Running the fluxgate secondary coil in the short-circuited mode, as described in Chap.7, offers a very simple input stage without the need for second harmonic band-pass filtering, and without the parasitic resonance effects caused by coil and cable capacitances, as often seen in the broad band signal of the unloaded voltage-coupled input stage. The B-field related signal is a series of short current impulses at the second harmonic of the excitation frequency with an amplitude proportional to B, and a feed-through signal at the odd harmonics independent of B. Figure 8.1 from Primdahl et al. (1994) shows the feed-through signal for $B = 0$ and the current impulses for $B = 20\,000$ nT with the feed-through removed. Compared to the current impulses the peak-to-peak feed-trough signal corresponds to about 3 400 nT for this typical 17 mm diameter amorphous metal ring-core sensor. The pure

8.1. DIGITAL DETECTION AND FEEDBACK FLUXGATE MAGNETOMETRY

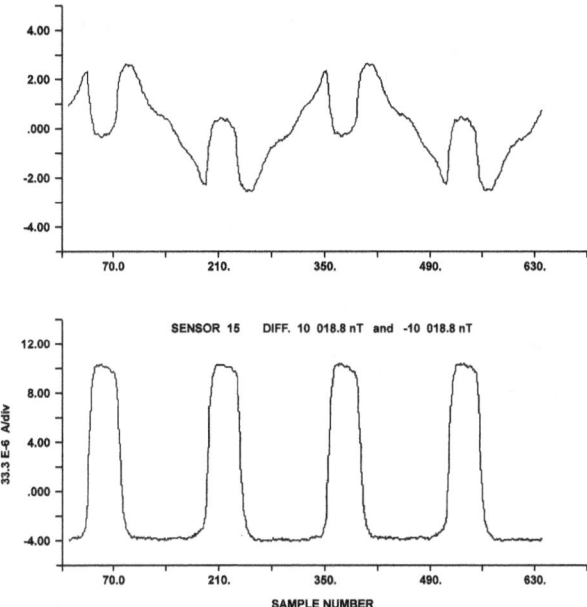

Figure 8.1: The top panel shows a sample of the zero field short-circuited fluxgate current recorded at the output of the current-to-voltage converter input stage. The vertical scale is 0.2 V/div. corresponding to 33.3 μA/div., and the horizontal scale is the sample number with 200 nsec between samples and covering 2 excitation periods. Bottom panel correspondingly shows the pure B-field signal, when the sensor is exposed to 20 000 nT calculated as the difference between the two actual signals including the feed-through shown in the top panel.

B-field current impulses are obtained by exposing the sensor to $\pm 10\,000$ nT and taking the difference between the two resulting wave forms. The signals were digitized with 8 bits and 320 samples per excitation period, and the excitation frequency was 15.625 kHz giving a sampling rate of 5 MHz.

8.1.3 The Impulse Integration Technique

An early version of a fluxgate magnetometer with digital detection and feedback system was introduced by Lühr (1980). This method takes advantage of the fact that the information about the external field is confined to the pulses of the sensor signal, as has been shown in the previous section. Lühr (1980) proved analytically and also demonstrated experimentally that the integral over the impulse is directly proportional to the ambient field strength. This fact was used to combine the fluxgate electronics directly with a dual-slope integrating analog-to-digital converter (A/D).

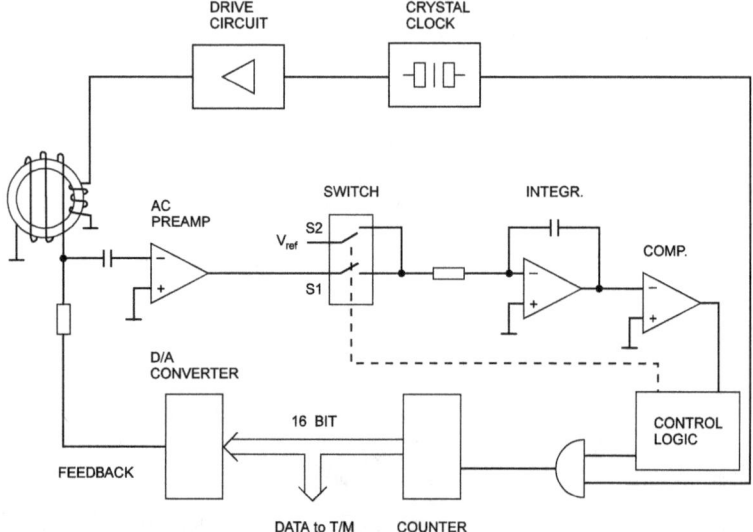

Figure 8.2: Schematic of the digital feedback magnetometer by Lühr (1980) used onboard the ROSE sounding rockets (Rose et al., 1990). Switch S1 time-gates the magnetic field sensitive part of the sensor output and opens for the integration of the impulse. Subsequently closing switch S2 the integrator is discharged and the time is found by the counter, the counter output is D/A converted for sensor feedback and is transferred to the TM as a measure of the external magnetic field.

Figure 8.2 from Lühr (1980) sketches the essential parts of a fluxgate employing the Impulse Integration Technique (IIT). The sensor output signal is feed through a preamplifier and more or less unconditioned connected to the input of the dual-slope A/D. Switch S1 connects the input signal to the integrator, whenever an impulse is expected. During the remaining time S1 is open. The gaps between the impulses are used to bring the output voltage of the integrator down again to zero. Depending on the sign of the input impulse a reference voltage with opposite sign is connected to the integrator until its output voltage reaches zero, as sensed by the adjacent comparator. The time it takes to bring down the integrator output is used to gate a number of oscillator cycles. An up/down counter changes its previous value according to the number of counts received and to the sign of the input impulse. The state of the counter represents the digital output word of the magnetometer. In order to improve the linearity of the system the digital word is also fed to a digital-to-analog converter (D/A) which facilitates a compensation of the ambient field in a feedback loop.

The principal features of this design are the digitization of the signal as close as possible to the sensor and at a speed as high as reasonable for a fluxgate system (twice the excitation frequency). The employed A/D incorporates a good part of the fluxgate electronic of conventional designs and all the advantages of a dual-slope integrating A/D like high linearity, immunity against high frequency noise and auto-zero-cycling improve the system.

Fluxgate magnetometers of the IIT design have successfully been flown on the ROSE sounding rockets (Rose et al., 1990). These instruments had a full-scale range of $\pm 55\,000$ nT digitized to 16 bits. The 8 kHz samples were averaged over 2 ms and transmitted to the ground. The resolution was determined by the digitization step of $1.7\ nT$ in this case. Power consumption (3W, incl. DC/DC converter) and mass (1.5 kg) are comparable with that of a conventional design including the digitization.

A prototype of a high resolution ground-based magnetometer employing the IIT design has been described by Ernst (1986). This instrument provides a 23 bit resolution at a sample rate of 128 s/s. The noise level determined in field applications ranges around 0.1 nT at a 1 Hz bandwidth and a $\pm 64\,000$ nT range. These results show that the IIT design exhibits performance characteristics at least as good as conventional fluxgate designs; but with the added advantage of an intrinsic digital output.

8.1.4 Real-Time Second Harmonic Digital Detection and Feedback Magnetometer

Auster et al. (1995) described a CPU based real-time operating digital detection and feedback magnetometer. They have developed hardware in which the near sensor signal is online digitized, its information about the B-field is evaluated, and a feedback field is generated. Figure 8.3 shows the hardware configuration.

The fluxgate ring-core excitation frequency is $f_0 = 6.5$ kHz and the preamplifier output contains (mainly) signals at f_0, $2f_0$ and $3f_0$ as indicated in Fig. 2 of Auster et al. (1995). The A/D samples the total signal at the Nyqvist frequency of the second harmonic, i.e at $4f_0 = 26$ kHz. As the sampling phase is adjusted to the exact phase of the expected B-field dependent second harmonic, then a complete recovery of the $2f_0$ signal is possible. The differences are calculated between the odd samples (corresponding to the

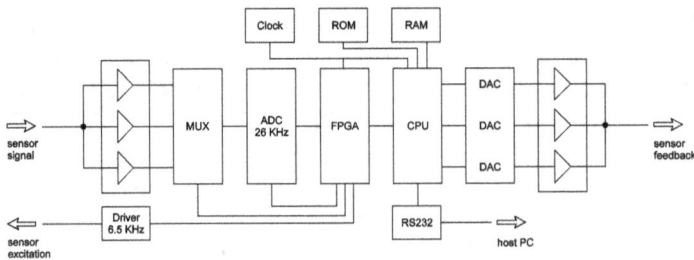

Figure 8.3: Hardware configuration of the second harmonic digital detection and feedback magnetometer by Auster et al. (1995).

positive extreme values of $2f_0$) and the even samples (the negative extreme values of $2f_0$). Digital low pass filtering is then applied to obtain the desired open loop bandwidth (e.g.10 Hz).

The simple, but fast arithmetics is performed in a separate FPGA (Field Programmable Gate Array) to relieve the CPU (Central Processing Unit). The CPU controls the feedback D/A's, serves as a data logger, and is responsible for the communication with the host computer.

The digital system is compared to two analogue versions of the magnetometer developed for MARS 94 and for Equator-S, and all three systems operate at the noise level of the ring core sensors (5 pT/$\sqrt{\text{Hz}}$ at 1 Hz). No information is given about the dynamic range of the magnetometers, but presumably they are all low field instruments intended for deep space missions. A 24 h run showed about 0.1 nT peak-to-peak difference fluctuations between the analogue and digital magnetometers (Auster et al., 1995, Figure 7).

The digital magnetometer is able to change the transfer function under S/W control by changing the feedback current and the mean value calculation (the low-pass open loop digital filter frequency), and magnetic field measurements can be done with a high level of automation. The data output rate and thus the power consumption can be controlled by changing the systems basic clock rate. From a commercial point of view, the digital magnetometer is a further step towards production of a low cost instrument because costs for manpower to align the filter will become dispensable.

8.1.5 Digital Cross-Correlation Detection and Feedback Fluxgate Magnetometer

Referring to Fig. (A) from Primdahl et al. (1994), the cross-correlation coefficient, r, between the total output signal $S(i)$ at the top and a reference signal $R(n)$ of the same shape as the expected B-field current impulses at the bottom will depend on the content of B-field signal in the total output, and on the phase relationship between the signal and the reference.

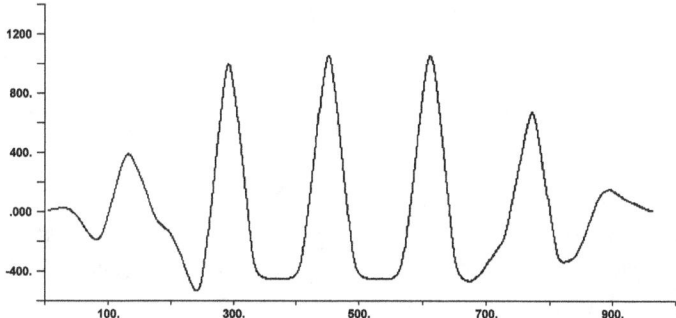

Figure 8.4: The cross-correlation function $r(i)$ between a 320 samples reference function $R(n)$ and a 640 samples signal $S(i)$. The vertical scale shows the (arbitrary) amplitude and the horizontal scale represents the phase relationship between $R(n)$ and $S(i)$. Owing to edge effects from the limited signal sample, $S(i)$, only i-values between 320 and 640 represent the true cross-correlation function.

Figure 8.4 shows the cross-correlation function, r, between 640 samples of the output signal (Fig. 8.1 top) for $B = 10\,000$ nT and a 320 ($= N$) samples reference function constructed from Fig. 8.1 bottom:

$$r\left(i - \frac{N}{2}\right) = \sum_n S(i - N + n) R(n), \qquad n = 1, N \qquad (8.1)$$

where $r(i - N/2)$ is a measure of the correlation between S and R at a time $N/2$ earlier than the last data sample $S(i)$ in the window. As time (i) changes, then r will change periodically reflecting the changing phase relationship between $R(n)$ and $S(i)$. Owing to the limited length of the sample $S(i)$, the points before 320 and after 640 are modified by edge effects. The cross-correlation function can be normalized, so that the extreme value directly gives the input magnetic field in nT, or in any other scaling needed for further calculations. All the samples shown here and in the following were real time digitized and stored for subsequent analysis, and the cross-correlation calculations were made off-line.

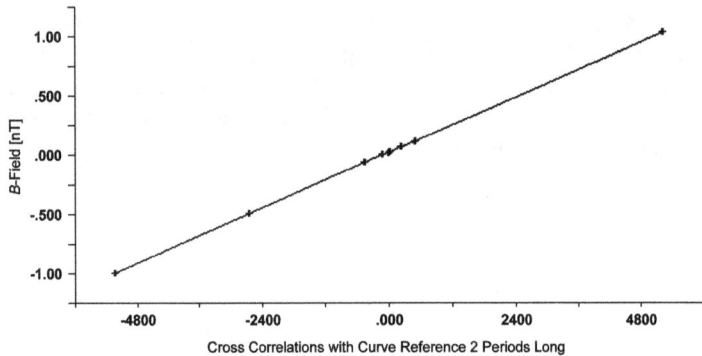

Figure 8.5: Input magnetic field plotted against the maximum cross-correlations for nine field settings. The linear relationship demonstrates that the cross-correlation is a reliable measure of the external magnetic field.

The cross-correlation extreme values respond to the input magnetic field as shown in Fig. 8.5. The scatter of the points about the best-fitted straight line shows a 12.5 nT standard deviation indicating that 15 625 magnetic field measurements per second can be made with this system with a noise level of about 12.5 nT root-mean-square (RMS). Averaging to 10 samples per second the calculated noise level is then expected to be 0.3 nT RMS.

A full Earth's field real-time digital detection and feedback magnetometer based on this principle is described by Piil-Henriksen et al. (1996). The single axis magnetometer is constructed around a commercially available DSP PC-board (PC/21K-040, Loughborough Sound Images) based on the Analog Devices ADSP21020 processor with the purpose of exploring and optimizing the capabilities of a digital system.

The ring-core fluxgate sensor has the same geometry and uses the same type of stress annealed amorphous magnetic material as the cores for the Danish Ørsted satellite magnetometer (Nielsen et al., 1995). The 17 mm diameter ring has 11 wraps of magnetic ribbon, and the secondary pickup coil has two sections of 295 turns each. The core noise is 14 $pTRMS$ in the band 60 mHz to 10 Hz corresponding to about 6 $pT/\sqrt{Hz}$ at 1 Hz.

The excitation driver circuit (Fig.8.6) is taken from the Ørsted magnetometer, and the broad band input stage (see Fig.8.7) converts the current impulses from the short-circuited sensor secondary coil to a voltage signal matching the processor board's 16 bit A/D input range of ±3 V. The oper-

8.1. DIGITAL DETECTION AND FEEDBACK FLUXGATE MAGNETOMETRY

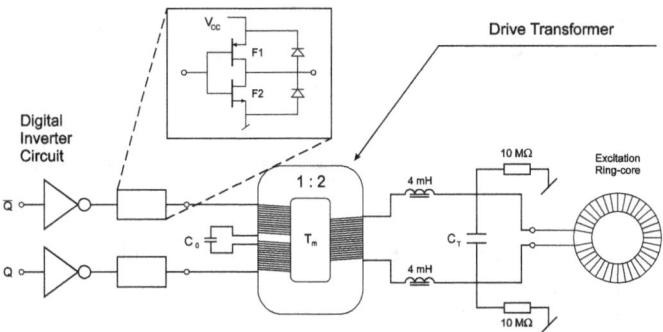

Figure 8.6: The excitation driver circuit. The balanced step-up transformer ensures symmetric drive current in the sensor drive cable with minimum overcoupling to the other cables. The tuning capacitor C_T is placed close to the sensor in order to avoid the large excitation current peaks in the drive cable.

ational amplifier is a current feedback circuit (AD846), which is inherently stable in this configuration as opposed to a voltage feedback type. The processor board's 16 bit D/A output (±3 V) directly drives the field nulling current through a separate compensation coil via the feedback resistor. This sets the ±52 000 nT full scale range (FSR) of the instrument.

The 2.1875 kHz excitation driver square wave (see Fig.8.7) is controlled by the DSP from the DSPLINK output via two D-flip-flops. It is locked to the 140 kHz A/D sampling rate resulting in exactly an integer 64 samples per excitation period with negligible timing jitter.

The output 16 bit D/A is controlled by the DSP to give one feedback update per excitation period, i.e.2.1875 ksamples/seconds, this is then averaged by filtering to 100 Hz or 10 Hz upper frequency limit for the noise measurements.

The output noise is71 pT RMS in the 0.25 Hz to 10 Hz bandwidth for this ±52 000 nT FSR instrument. Generally a ratio of about −123 dB exists for this digital instrument between the FSR and the RMS-noise, which indicates that the D/A and the feedback circuit is responsible for the instrument output noise exceeding the core noise.

8.1.6 The Detection and Control Algorithms

Figure 8.1, bottom panel, shows that the B-field current impulses closely resemble a series of box-shaped impulses. For that reason, and also because the box-shaped reference is simpler to implement, this was chosen rather than a reference of exactly the same wave-form as the B-field impulses. The box-reference was, of course, shaped to have maximum cross-correlation with the B-field signal.

The equivalence between this peak cross-correlation and the integral over the voltage output impulse used in the IIT-method can be shown mathematically. Both are related to the core flux changes induced during the excitation cycle by the external field, as discussed in Primdahl et al. (1991).

The cross-correlation function $r(k)$ between the digital signal $S(i)$ and the reference $R(n)$ is defined by:

$$r(k) = \frac{1}{N} \sum_n S(k+n)\, R(n)\,, \qquad n = 0, N-1 \qquad (8.2)$$

Here N is the length of the considered window, which is equal to one excitation period.

In the time domain this is a series of multiplications followed by a summation, which is equivalent to a time-convolution or filtering. It may be implemented as a finite impulse response (FIR) filter with filter coefficients a_p matching the (box-shaped) reference:

$$y(n) = \sum_p a_p x(n-p)\,, \qquad p = 0, N-1 \qquad (8.3)$$

where $x(n)$ are the input values to the filter, and each output value $y(n)$ represents a weighted average of the preceding N samples of the input. The Fourier's spectrum of the FIR-filter, having zero's at the odd harmonics of the excitation frequency, is indicated in the DSP box in Fig.8.7.

The FIR-filter output is a periodic function resembling the (middle part of the) wave form in Fig.8.4, and the extreme values of the cross-correlations are picked out by multiplying a spike wave form onto the FIR-filter output, where the spikes are synchronized to the cross-correlation peaks (see the DSP-box in Fig.8.7). After multiplication with the amplification parameter A_m the values are summed in the integrator register once every excitation period, and the integrator updates the D/A at the same rate, i.e. every $0.45714\ ms$.

The digital feedback loop is equivalent to a first order inherently stable analogue loop, and the overall magnetometer frequency response is controlled by the total loop gain and regulated by the amplification factor A_m. The open loop frequency response approaches half the excitation frequency according to the Nyqvist theorem.

8.1.7 The Astrid-2 Satellite Magnetometer

Based on the laboratory experiments described above, a digital satellite magnetometer has been developed for the Swedish 30 kg Astrid-2 micro satellite as a joint collaboration between Danish, Swedish and Finnish groups.

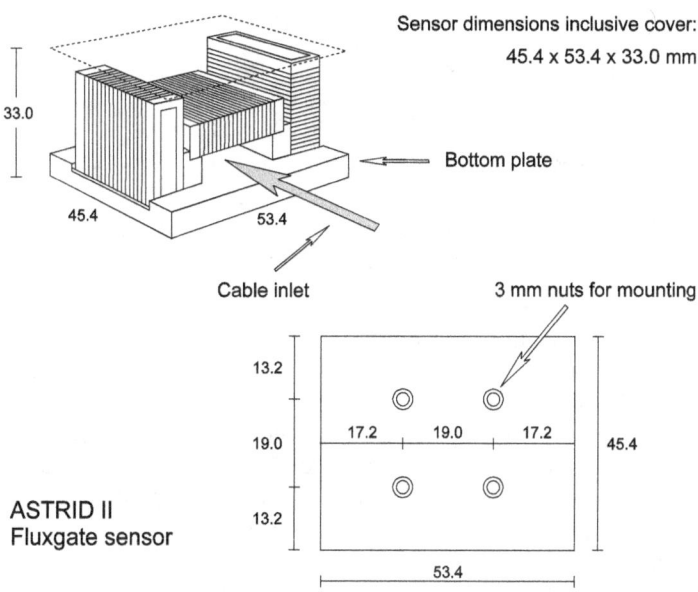

Figure 8.8: The Astrid-2 satellite tri-axial fluxgate sensor. The sensor weight is 130 g including cover but exclusive cables.

The miniature tri-axial sensor is based on the development of the Ørsted satellite magnetometer (Nielsen et al., 1995). The mechanical structure (see Fig. 8.8) is made of MACOR, a highly stable machinable glass material, and a thermistor in the sensor measures the temperature for correction for the linear thermal dependencies of the sensor calibration matrix coefficients. The cross-couplings between the individual fluxgates, and the transverse field dependencies are highly linear for these amorphous metal ring-core sensors, and all are included in the sensor response matrix coefficients. Low feed-through and high linearity is obtained by optimizing the core ribbon stress annealing treatment (Nielsen et al., 1995).

Figure 8.9 shows a block diagram of one component of the basic magnetometer. The 17 mm diameter ring-core fluxgate is magnetized by an 8 kHz excitation AC-current derived from the common external timing controller. The short-circuited current output from the secondary coil is conditioned

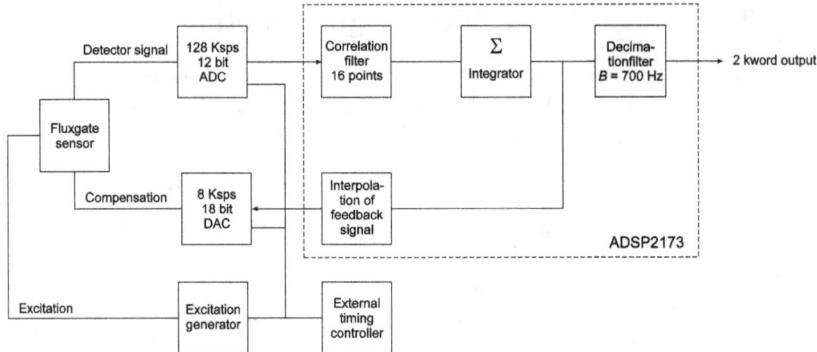

Figure 8.9: Block diagram of one component of the DSP based Astrid digital magnetometer. Dotted line indicates the DSP functions.

for the 12 bits 128 ksamples/second A/D, which provides 16 samples per excitation period in exact phase synchronization with the excitation current.

The Analog Devices ADSP2173 DSP handles the digital input correlation filter, the integration, the feedback signal interpolation, and the 700 Hz low pass output digital filter; and it transfers 2 kwords/second to the telemetry interface. The feedback D/A is an 18 bits audio type with a25 $ppm/$ temperature coefficient, and it controls a constant current source for driving the compensation current through the sensor secondary coil. The D/A update rate can be 8 kHz or 16 kHz corresponding to f_0 or to $2f_0$.

The magnetometer full scale range is ±65 000 nT with a noise level of 40 pT RMS in the range 50 mHz to 10 Hz. This is more than the 14 pT RMS sensor noise measured in the same frequency band and in an open loop configuration; but it is highly adequate for the magnetic mission onboard Astrid-2.

Power consumption is 1.9 W, weight of the electronics box is about 750 g, the sensor weighs 130 g, and sensor cables are 4 twisted shielded pairs plus one twisted pair for the thermistor weighing in total about 22 g/m.

8.1.8 Discussion and Conclusions

The instrument developments reported on here demonstrate the feasibility of replacing virtually all analog signal handling in the classical second harmonic fluxgate magnetometer by digital signal processing.

The Impulse Integration Technique (IIT) performs the cross-correlation using analogue electronics; but the digital feedback loop function is implemented using a simple up-down counter controlled by letting in effect the sensor compare the external field with the feedback field.

It is a slight disadvantage of the IIT that the switching time of S1 cannot be adjusted to the appearance of the input impulse by remote control, and the same is true for the transfer function of the magnetometer. This is an inherent price payed for having such a simple circuit without a CPU.

The second harmonic digital detection circuit of Auster et al. (1995) uses synchronous sampling and second harmonic detection by an FPGA, and the digital feedback loop is controlled by a CPU. The detector phasing and the transfer characteristic may then be remotely controlled, and so can the clock frequency, the data taking rate and thus the power consumption.

The cross-correlation digital detection system by Piil-Henriksen et al., (1996) extracts the the maximum available information about the magnetic field from all the significant even harmonics of the signal.

Despite the intrinsic complexity of a DSP with supporting circuits, the total number of IC's and discrete components may actually decrease compared to that of the conventional analogue circuits, thereby offering an increased overall reliability.

Advantages of the systems are expected to be the avoidance of overcoupling of an analogue reference to the magnetometer input, where it may be detected as a DC-offset, and the implementation of a digital correlator with far better rejection of noise and odd-harmonics signals than that of a corresponding analogue synchronous phase detector. The declared goal of constructing a full Earth's field FSR instrument with only the sensor noise has not yet been fully achieved; but results compare favorably with those of the Ørsted analogue magnetometer, and improvements are being investigated.

Besides performing all the basic signal conditioning functions the presence of a DSP under software control also opens for added instrument flexibility and the introduction of autonomy for instrument adaptation to changing external conditions. Diagnostics and error handling routines may be included in the software, and so may self calibration, offset control and linearity tests.

References chapter 8

1. Ernst, A., Aufbau und Erprobung eines hochauflösenden Breitbandmagnetometers, Dipl.-Arb., Inst. f. Geophys. Meteorol., Techn. Univ., Braunschweig, 1986.

2. Lühr, H., Das Impulsintegrationsverfahren: Ein Saturationskernmagnetometer mit digitaler Messwertdarstellung, Diss., Techn. Univ. Braunschweig, 1980.

3. McPherron, R.L., P.J. Coleman, Jr. and R.C. Snare, ATS-6 UCLA Fluxgate Magnetometer, IEEE Trans. Aerospace and Electron. Systems, AES-11, 1110-1116, 1975.

4. Rose, G., K. Rinnert, K. Schlegel, E. Neske, A. Friker, F.-J. L!bken, A. Steinweg, D. Krankowsky, P. Lämmerzahl, K. Mausberger, B. Anweiler, H. Lühr, W. Oelschlägel, G. Dehmel, J.Warnecke, H. Kohl, E. Nielsen, Rocket and Scatter Experiments (ROSE): Final report on the scientific aspect, Max-Planck-Institut für Aeronomie, MPAE-W-40-90-18, 1990.

Chapter 9

Magnetometer Ground Calibration

HolgerKügler

*IABG – Industrieanlagen Betriebsgesellschaft mbH,
München/Ottobrunn*

9.1 Calibration Principle

The introduced calibration method is based on measurements in a triaxial coil facility (like the Braunbek coil facility in Braunschweig or the big IABG facility at Ottobrunn/Munich). The instrument to be calibrated is placed in the center and exposed to several well defined (calibrated by absolute instruments) magnetic field configurations[1]. These well known magnetic field configurations are used as reference. The magnetometer answer to these field configurations their strength and orientation as well as the ambient temperature are measured. The process of a complete calibration is illustrated in Fig.9.1 schematic. All different field configurations are generated and controlled by the 'Field Configuration Control' within the facility. The 'answer' of the instrument to this field configurations and the sampled environment parameter like the temperature within the facility are measured by the 'Data Acquisition System' and judged by the 'Analysis'.

The qualified data is used to provide correction parameter for a further improvement of the instrument. In case that the instrument is used as a 'reference' it is even possible to increase the knowledge of the facility used

---

[1]The strength and orientation of the applied magnetic field vector is known.

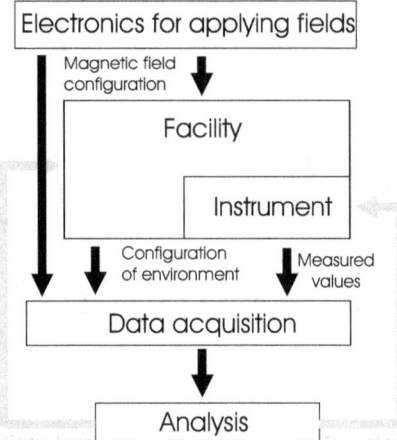

Figure 9.1: Process of a calibration measurement (schematic)

for the calibration. In a second step this knowledge can be used to increase the accuracy of the calibration process itself. The schematic process of a calibration measurement as described in Fig.9.1 schematic looks like this:

1. Placing the instrument within the facility.

2. Applying a magnetic field configuration by the coil facility.

3. Wait until the the magnetic field configuration within the facility is settled.

4. Wait until the instrument to be calibrated is settled and stable.

5. Reading of the facility magnetic field configuration values and of the instrument values and combining these readings to a data vector.

6. Repeating these steps 2 until a sufficient number of data vectors has been stored.

These vectors called calibration vectors are used as basis of the following analysis.

9.2 Co-Ordinate Systems

The above procedure requires the introduction of several co-ordinate systems. The magnetic fields inside the facility are described by a Cartesian co–ordinate system with axes named U, V, W and their related unity vectors $\mathbf{e_u}$, $\mathbf{e_v}$, $\mathbf{e_w}$. The co-ordinate system used to describe the magnetic fields is named X, Y, Z and its components are x, y, z. Typically this co-ordinate system is not orthogonal and is not well aligned with respect to the other co-ordinate systems.[2]

9.3 Simple Example

The following simple example presents the calibration process of a magnetometer. The example is based on the following assumptions: A ring core fluxgate magnetometer for space application is calibrated. It shall be verified whether the instrument fits the magnetic requirements.

Following the theory of Flux gate-Magnetometers [8] a linear transfer function is expected[3]. In order to check this it is necessary to create and to measure well defined stable magnetic field configurations. In this example it is assumed that the facility is ideal and described by a right handed Cartesian co-ordinate system named U, V, W. The magnetic field (**b**) measured in this co-ordinate system is given by the components u, v, w. The co-ordinate system of the magnetometer (Instr) itself uses the axes X, Y, Z. Its components are named x, y, z. Using this nomenclature and assuming a linear transfer function for transformation leads to the following formulas:

$$b_{\mathsf{Instr},x} = a_{x,1}b_u + a_{x,2}b_v + a_{x,3}b_w, \tag{9.1}$$

$$b_{\mathsf{Instr},y} = a_{y,1}b_u + a_{y,2}b_v + a_{y,3}b_w, \tag{9.2}$$

$$b_{\mathsf{Instr},z} = a_{z,1}b_u + a_{z,2}b_v + a_{z,3}b_w \tag{9.3}$$

or by using vectors

$$\mathbf{b}_{\mathsf{Instr}} = \underline{\underline{A}}\, \mathbf{b}_{\mathsf{Labor}}. \tag{9.4}$$

The calculated transfer coefficients $a_{\ldots,1}$, $a_{\ldots,2}$ and $a_{\ldots,3}$ are related to the test set-up.

---

[2]In addition the strength and orientation of a magnetic field vector is often depending on the mechanical position of the observation point to be used. Therefore it is necessary to introduce a second Cartesian co-ordinate system named U_L, V_L, W_L and its unity vectors $\mathbf{e_{u_L}}$, $\mathbf{e_{v_L}}$, $\mathbf{e_{w_L}}$. This co–ordinate system is co-aligned to the UVW–system.
A second co–ordinate system is typically related to the instrument (Instr). This mechanical co-ordinate system is used to describe the geometrical position and orientation of any mechanical part of the instrument. This co-ordinate system is a Cartesian one with axes named X_{mech}, Y_{mech}, Z_{mech} and its components $\mathbf{e_{x_{\mathsf{mech}}}}$, $\mathbf{e_{y_{\mathsf{mech}}}}$, $\mathbf{e_{z_{\mathsf{mech}}}}$. This co-ordinate system is different from the one used to describe the magnetic fields measured by the instrument. Both mechanical co-ordinate systems are not mentioned furthermore
[3]neglecting possible non linearities of the magnetometer electronics.

The transfer coefficients of the matrix $\underline{\underline{A}}$ can be calculated by the variation of the magnetic field configuration b_u, b_v, b_w within the facility and the the magnetometer answer. Calculating the transfer coefficients $a_{x,1}$, $a_{x,2}$, $a_{x,3}$ by using N measurements leads to:

$$\underbrace{\begin{pmatrix} b_{u,1} & b_{v,1} & b_{w,1} \\ b_{u,2} & b_{v,2} & b_{w,2} \\ \vdots & \vdots & \vdots \\ b_{u,N} & b_{v,N} & b_{w,N} \end{pmatrix}}_{\underline{\underline{B}}} \underbrace{\begin{pmatrix} a_{x,1} \\ a_{x,2} \\ a_{x,3} \end{pmatrix}}_{\mathbf{x}} = \underbrace{\begin{pmatrix} b_{x,1} \\ b_{x,2} \\ \vdots \\ b_{x,N} \end{pmatrix}}_{\mathbf{m}}. \quad (9.5)$$

The coefficients $a_{x,1}$, $a_{x,2}$, $a_{x,3}$ of the vector $\mathbf{x}$ can be calculated using a least mean square fit. This can be done by introducing the quadratic distance function $Q(\mathbf{x})$:

$$Q(\mathbf{x}) = (\underline{\underline{B}}\,\mathbf{x} - \mathbf{m})^T (\underline{\underline{B}}\,\mathbf{x} - \mathbf{m}) = \mathbf{x}^T \underline{\underline{B}}^T \underline{\underline{B}}\,\mathbf{x} - \mathbf{m}^T \underline{\underline{B}}\,\mathbf{x} - \mathbf{x}^T \underline{\underline{B}}^T \mathbf{m} + \mathbf{m}^T \mathbf{m}. \quad (9.6)$$

To minimize $Q(\mathbf{x})$ it is necessary that all deviations of x_i are zero. It is:

$$\nabla Q(\mathbf{x}) = 0. \quad (9.7)$$

Therefore equation 9.6 transforms to:

$$\begin{aligned} \nabla Q(\mathbf{x}) &= \nabla (\mathbf{x}^T \underline{\underline{B}}^T \underline{\underline{B}}\,\mathbf{x} - \mathbf{m}^T \underline{\underline{B}}\,\mathbf{x} - \mathbf{x}^T \underline{\underline{B}}^T \mathbf{m} + \mathbf{m}^T \mathbf{m}) & (9.8) \\ &= 2\,\underline{\underline{B}}^T \underline{\underline{B}}\,\mathbf{x} - 2\,\underline{\underline{B}}^T \mathbf{m} \stackrel{!}{=} 0. & (9.9) \end{aligned}$$

With

$$\underline{\underline{B}}^T \underline{\underline{B}}\,\mathbf{x} = \underline{\underline{B}}^T \mathbf{m}. \quad (9.10)$$

In case that matrix $\underline{\underline{B}}$ has the full Rang of its columns it is possible to invert matrix $\underline{\underline{B}}^T \underline{\underline{B}}$ with the resolution vector:

$$\mathbf{x} = (\underline{\underline{B}}^T \underline{\underline{B}})^{-1} \underline{\underline{B}}^T \mathbf{m}. \quad (9.11)$$

The parameter in equation (9.2) and equation (9.3) can be calculated in the same way. Generating and measuring at least three independent magnetic vectors within the facility leads to a solution of the equation system. But the quality of the fit is increasing with the number of magnetic vectors taken. Furthermore it is essential that the space within the magnetic co-ordinate system is covered equidistant by many measuring points. The so calculated coefficients represent the best fit of the linear model. In addition section 9.7

9.3. SIMPLE EXAMPLE

describes how to solve this equation system for $N > 300$ without explicit calculation of the formula $(\underline{\underline{B}}^T \underline{\underline{B}})^{-1}$.

The number of measuring steps within a manual driven facility is often very limited. Furthermore it is often necessary to eliminate writing errors of the operator. Therefore a special sequence so called 'on-axis' measurements is used. During this sequence the magnetic field is applied only along the co-ordinate axes of the facility. Taking a manual handling into account the following sequence can be used:

1. Generate zero magnetic field on all axes U,V,W

2. Generate magnetic fields in direction of the U–axis. Beginning with a field Min_u up to Max_u with a step width of s_u. No additional field is applied on the V and W axes.

3. Generate zero magnetic field on all axes

4. Generate magnetic fields in direction of the V–axis. Beginning with a field Min_v up to Max_v with a step width of s_v. No additional field is applied on the U and W axis.

5. Generate zero magnetic field on all axes

6. Generate magnetic fields in direction of the W–axis. Beginning with a field Min_w up to Max_w with a step width of s_w. No additional field is applied on the U and V axes.

7. Generate zero magnetic field on all axes

Figure 9.2 indicates that one always tries to align the instrument with respect to the facility axes. 'Try', because there are often reasons which don't allow this. It is remarkable that the answer of the instrument is always negative to positive field values of the facility. In a first approximation the following transfer matrix is valid:

$$\mathbf{b}_{\text{Instr}} \approx \begin{pmatrix} -1 & 0 & 0 \\ 0 & -1 & 0 \\ 0 & 0 & -1 \end{pmatrix} \mathbf{b}_{\text{Facility}} = \underline{\underline{A}}_{\text{approx}} \mathbf{b}_{\text{Facility}} \quad (9.12)$$

The determinate of this matrix is -1. Therefore the instrument is related to a left handed co–ordinate system. It makes sense to multiply all field vectors $\mathbf{b}_{\text{Facility}}$ by matrix $\underline{\underline{A}}_{\text{approx}}$ to make the input fields generated by the facility direct comparable to output fields measured by the instrument.

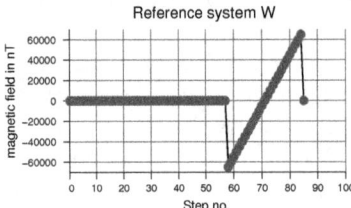 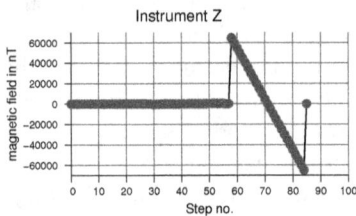

Figure 9.2: This figure illustrates a sequence of magnetic fields of an 'on-axis' measurement. The magnetic field generated within the facility is plotted versus the calibration step number on the left side of the figure. The right side of the figure shows the 'answer' of the instrument for these magnetic field configurations. The field values of each calibration step are summarized as test vector. The answer of the instrument consists of the answer of each single sensor axis. Here only Z is shown.

This is shown on the left column of figure 9.3. The columns in the middle (difference A) show the difference after subtracting the measured values from the input values. The difference A $= \mathbf{b}_{\text{Facility}} - \mathbf{b}_{\text{Instr}}$. This figure compared with figure 9.2 shows that the maximal difference is in the order of 800 nT / 65000 nT $\approx$ 1.2 percent. The result of multiplying $\underline{\underline{A}}_{\text{approx}}$ with $\mathbf{b}_{\text{Facility}}$ is called model vector $\mathbf{b}_{\text{Model}}$. Taking all model vectors and running a least mean square fit with linear coefficients leads to a matrix A_{Fit}. The residual fields shown in the difference B on the right side of the figure are calculated by

$$\mathbf{b}_{\text{Residual}} = \underline{\underline{A}}_{\text{Fit}} \mathbf{b}_{\text{Model}} - \mathbf{b}_{\text{Instr}}. \qquad (9.13)$$

Examining these residual fields it is possible to decide whether an instrument fits its requirements or not. In case that the residual fields are greater than the specification it is for sure that the instrument does not fit the specification or a 'wrong' model is used to characterize the instrument. If model and instrument fit well it can be assumed that the residual fields show no 'structure'plotted versus any input parameter. If one discover any structures it is for sure that the model does not fit the measured values well and/or that other systematical effects have to be taken into account.

Looking at this example the residual fields of the X–component show a non linear behavior of magnetic fields applied to the V–axis of the facility. This axis is perpendicular to the X-component and therefore should have no impact on it. In the literature this effect is called magnetic **non linear cross coupling** [1]. This effect can be seen at several Ring core-Flux

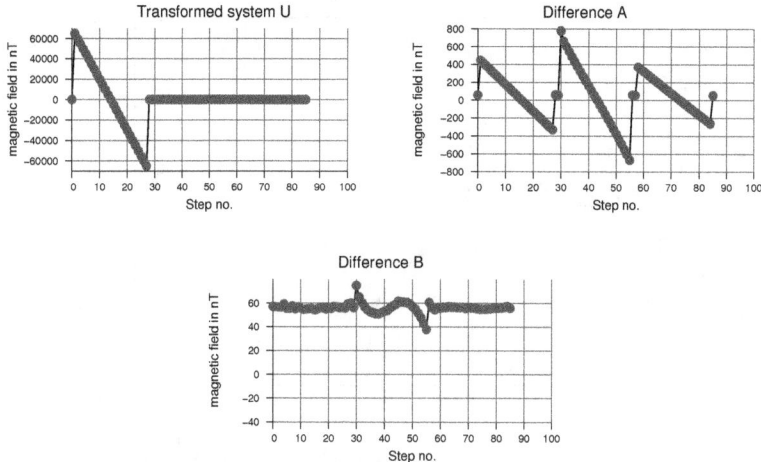

Figure 9.3: fitting of different models

gate-Magnetometer that are used in 'strong' magnetic background field. This non linear effect is in the order of 0.06 percent with respect to the maximum of the magnetic field set up in the facility. On the other hand this implies that the linearity of the tested instrument is better than 0.1 percent or corresponds to an accuracy of 8 nT in front of a background magnetic field of 65000 nT.

9.4 Magnetic Environment Requirements During Calibration

An accuracy and reproducibility during measurements of better than 1 nT within the facility is especially requested when a sensitive magnetometer is calibrated. Therefore the ambient local magnetic field has to be controlled during measurements, the following aspects have to be taken into account:

The ambient magnetic field is variable in strength and direction depending on location and time. An impression of this behavior is given in Figure 9.4. The daily variation of the total intensity of the magnetic vector on a magnetic quiet day is shown in this figure. The magnetometer used during these measurements was placed at a location where the effect of magnetic fields created by civilization has been minimized.

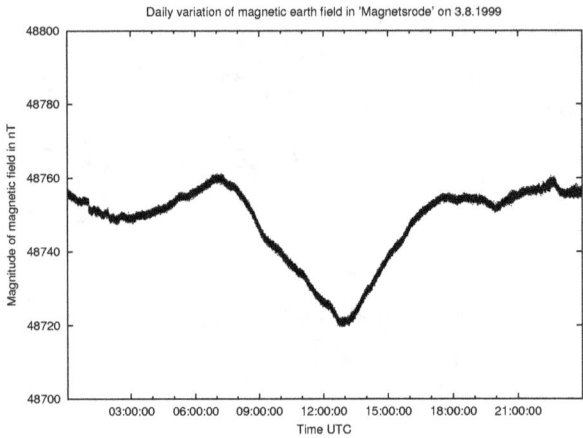

Figure 9.4: Variation of the magnetic field of Earth on a magnetic 'quiet' day

A way to eliminate ambient magnetic fields is a dynamic compensation of the ambient magnetic field.

9.5 Parameters of a Linear Instrument

The model of a linear model of an instrument can be used to determine its orientation and the sensitivity with respect to the so called facility co-ordinate system. This kind of model expects a linear transfer function between facility co-ordinate system and the co-ordinate system of the instrument.

Figure 9.5 illustrates this idea by using a geometrical approach. It is assumed that for example a magnetic field is generated in direction of the U–axis. The length of this vector symbolizes the strength of the magnetic field. A sensor S measures the magnetic induction along the direction of the arrow. The sensor itself discovers only that part of the magnetic induction pointing in U-direction which is projected in the direction of the sensor S.

Assuming a linear amplification in the electronics the output of the magnetometer X–channel is:

$$b_{\mathsf{Instr,x}} = e_1 \cos(\angle[S, U]) b_{\mathsf{u}}. \tag{9.14}$$

9.5. PARAMETERS OF A LINEAR INSTRUMENT

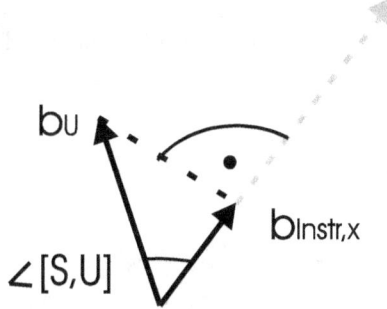

Figure 9.5: Geometrical interpretation of a magnetic induction

This results in the following output of one magnetometer component:

$$b_{\text{Instr},x} =$$
$$= e_1 \cos(\angle[S,U])b_u + e_1 \cos(\angle[S,V])b_v + e_1 \cos(\angle[S,W])b_w \quad (9.15)$$
$$= e_1 \left(\cos(\angle[S,U])b_u + \cos(\angle[S,V])b_v + \cos(\angle[S,W])b_w\right). \quad (9.16)$$

If the sensor is not moved during measurement $\cos(\angle[S,U])$, $\cos(\angle[S,V])$ and $\cos(\angle[S,W])$ are constant. Therefore it is possible to compare the coefficients of Eq.(9.16) to receive a linear transformation

$$b_{\text{Instr},x} = l_1 b_u + l_2 b_v + l_3 b_w \quad (9.17)$$

with

$$l_1 = e_1 \cos(\angle[S,U]), \quad (9.18)$$
$$l_2 = e_1 \cos(\angle[S,V]), \quad (9.19)$$
$$l_3 = e_1 \cos(\angle[S,W]) \quad (9.20)$$

Assuming that the co-ordinate system of the facility is orthogonal and normalized and taking into account

$$1 = \cos(\angle[S,U])^2 + \cos(\angle[S,V])^2 + \cos(\angle[S,W])^2$$

we get

$$e_1 = \pm\sqrt{l_1^2 + l_2^2 + l_3^2}, \quad (9.21)$$
$$\cos(\angle[S,U]) = \frac{l_1}{e_1}, \quad (9.22)$$
$$\cos(\angle[S,V]) = \frac{l_2}{e_1}, \quad (9.23)$$
$$\cos(\angle[S,W]) = \frac{l_3}{e_1}. \quad (9.24)$$

Using a positive sign in Eq.(9.21) it is possible to describe the orientation and sensitivity. The coefficients l_1, l_2, l_3 can be calculated directly using a least mean square fit. For further information see section 9.7.

The linear model of a single sensor can be extended directly to a three axis sensor. A possible interaction of the sensor axes is not taken into account.

$$b_{\mathsf{Instr},x} = e_1 \cos(\angle[S_x, U])b_u + e_1 \cos(\angle[S_x, V])b_v + e_1 \cos(\angle[S_x, W])b_w, \quad (9.25)$$

$$b_{\mathsf{Instr},y} = e_2 \cos(\angle[S_y, U])b_u + e_2 \cos(\angle[S_y, V])b_v + e_2 \cos(\angle[S_y, W])b_w, \quad (9.26)$$

$$b_{\mathsf{Instr},z} = e_3 \cos(\angle[S_z, U])b_u + e_3 \cos(\angle[S_z, V])b_v + e_3 \cos(\angle[S_z, W])b_w \quad (9.27)$$

or as vectors

$$\mathbf{b}_{\mathsf{Instr}} = \underline{\underline{M}}_1 \, \mathbf{b}_{\mathsf{Facility}} \qquad (9.28)$$

with

$$\underline{\underline{M}}_1 = \begin{pmatrix} e_1 \cos(\angle[S_x, U]) & e_1 \cos(\angle[S_x, V]) & e_1 \cos(\angle[S_x, W]) \\ e_2 \cos(\angle[S_y, U]) & e_2 \cos(\angle[S_y, V]) & e_2 \cos(\angle[S_y, W]) \\ e_3 \cos(\angle[S_z, U]) & e_3 \cos(\angle[S_z, V]) & e_3 \cos(\angle[S_z, W]) \end{pmatrix}. \qquad (9.29)$$

Again it is possible to determine the sensitivities and orientation of the three sensor components with respect to the facility co-ordinate system. In addition it is possible to calculate the inverse transformation by exchanging input and output values. Then we get:

$$\mathbf{b}_{\mathsf{Labor}} = \underline{\underline{M}}_2 \, \mathbf{b}_{\mathsf{Instr}}. \qquad (9.30)$$

This approach including the calculation of the inverse coefficients allows a correction of the values measured during flight. Matrix $\underline{\underline{M}}_2$ is not the inverse matrix of $\underline{\underline{M}}_2$ due to the method of least mean square fit used.

In the example from section 9.3 the matrix $\underline{\underline{M}}_1$ is calculated by

$$\underline{\underline{M}}_1 = \begin{pmatrix} 0.99404 & -0.010838 & -0.0048864 \\ 0.0098803 & 0.98913 & -0.0084339 \\ -0.0010818 & -0.0065943 & 0.99732 \end{pmatrix}. \qquad (9.31)$$

Matrix $\underline{\underline{M}}_2$ describing the inverse transformation shows

$$\underline{\underline{M}}_2 = \begin{pmatrix} 1.0057 & 0.010744 & 0.0050266 \\ -0.010009 & 1.0117 & 0.0093979 \\ 0.00096227 & 0.0070052 & 1.0028 \end{pmatrix}. \qquad (9.32)$$

9.5. PARAMETERS OF A LINEAR INSTRUMENT

$\underline{\underline{M}}_1$ to be the inverse of $\underline{\underline{M}}_2$ has to fulfill

$$\underline{\underline{M}}_1 \underline{\underline{M}}_2 \stackrel{!}{=} \underline{\underline{I}} \qquad (9.33)$$

with Identity matrix $\underline{\underline{I}}$. The product of $\underline{\underline{M}}_1$ and $\underline{\underline{M}}_2$ looks like

$$\underline{\underline{M}}_1 \underline{\underline{M}}_2 = \begin{pmatrix} 0.999818 & -0.000306 & 0.000008 \\ 0.000057 & 1.000749 & 0.000889 \\ 0.002111 & 0.000306 & 1.000049 \end{pmatrix}. \qquad (9.34)$$

Based on matrix $\underline{\underline{M}}_2$ it is possible to estimate nine parameter of the instrument:

- (3x) sensitivity,

- (3x) internal alignment of the instrument,

- (3x) direction of the magnetic axes of the instrument with respect to the coil facility.

Furthermore it is possible to determine

- a matrix to orthogonalize the data of the instrument,

- and the rotation of the sensor with respect to the facility

using the parameter of the instrument.

$\underline{\underline{M}}_2$ of the model in Eq.(9.30) includes partly components describing the test set-up during the calibration showing the rotation of the sensor with respect to the facility.

After calibration the sensor is typically used in a different mechanical position and orientation as during calibration. Therefore this specific part concerning the position and orientation has to be separated for elimination. In the following a method is suggested to do this:

$\underline{\underline{M}}_2$ is separated into three matrices

$$\underline{\underline{M}}_2 = \underline{\underline{D}}_{Euler} \underline{\underline{O}} \underline{\underline{E}}. \qquad (9.35)$$

The transformation of the vectors measured are separated into three steps:

1. The values measured are first corrected by their sensitivity ($\underline{\underline{E}}$).

2. Second an orthogonalization using $\underline{\underline{O}}$

3. $\underline{\underline{D}}_{Euler}$ is turning the vectors into coil facility direction.

This gives an algorithm for correcting the measured values:

$$\mathbf{b}_{corrected} = \underline{\underline{O}}\,\underline{\underline{E}}\,\mathbf{b}_{Instr}. \qquad (9.36)$$

Matrix $\underline{\underline{D}}_{Euler}$ is not used here, it is part of the calibration set-up and therefore can be chosen separately.

$\underline{\underline{M_2}}$ is separated as follows:

Step 1: Separation of sensitivity matrix

Assuming that matrix $\underline{\underline{E}}$ is describing a linear transformation the sensitivity of the instrument is linear:

$$\underline{\underline{M_2}} = \begin{pmatrix} \frac{m_{11}}{E'_x} & \frac{m_{12}}{E'_y} & \frac{m_{13}}{E'_z} \\ \frac{m_{21}}{E'_x} & \frac{m_{22}}{E'_y} & \frac{m_{23}}{E'_z} \\ \frac{m_{31}}{E'_x} & \frac{m_{32}}{E'_y} & \frac{m_{33}}{E'_z} \end{pmatrix} \begin{pmatrix} E'_x & 0 & 0 \\ 0 & E'_y & 0 \\ 0 & 0 & E'_z \end{pmatrix} = \underline{\underline{M_2}}'\,\underline{\underline{E}}, \qquad (9.37)$$

with

$$E'_x = \sqrt{m_{11}^2 + m_{21}^2 + m_{31}^2}, \qquad (9.38)$$

$$E'_y = \sqrt{m_{12}^2 + m_{22}^2 + m_{32}^2}, \qquad (9.39)$$

$$E'_z = \sqrt{m_{13}^2 + m_{23}^2 + m_{33}^2} \qquad (9.40)$$

The length of E'_x, E'_y, E'_z is given by normalizing the unity vectors in facility co-ordinates. The internal alignment of the sensor axes with respect to each other is derived from the alignment of each single sensor axis with respect to the facility co-ordinate system. It can be calculated using the scalar product between different axes and is. regarding the single sensor axis this is:

$$\cos\xi_{xy} = \frac{1}{E'_x E'_y}[m_{11}m_{12} + m_{21}m_{22} + m_{31}m_{32}], \qquad (9.41)$$

$$\cos\xi_{yz} = \frac{1}{E'_y E'_z}[m_{12}m_{13} + m_{22}m_{23} + m_{32}m_{33}], \qquad (9.42)$$

$$\cos\xi_{xz} = \frac{1}{E'_x E'_z}[m_{11}m_{13} + m_{21}m_{23} + m_{31}m_{33}]. \qquad (9.43)$$

$$\underline{\underline{M_2}}' = \underline{\underline{D}}_{Euler}\,\underline{\underline{O}}. \qquad (9.44)$$

9.5. PARAMETERS OF A LINEAR INSTRUMENT

The development of matrix $\underline{\underline{O}}$ is done assuming that the X-axis of the instrument and the U-axis of the facility co-ordinate system are identical $\mathbf{e}_1$,(see fig.below). Furthermore it is assumed that the Y-axis of the instrument is located in the UV-plane $\mathbf{e}_2$. The orientation of the Z-axis is given by two angles between the U-axis and the V-axis respectively the W-axis $\mathbf{e}_3$. Inserting $\mathbf{e}_1$–$\mathbf{e}_3$ into the transformation matrix $\underline{\underline{O}}$ we get

$$\underline{\underline{O}} = \begin{pmatrix} 1 & \sin\beta & \sin\gamma \\ 0 & \cos\beta & \cos\gamma \sin\eta \\ 0 & 0 & \cos\gamma \cos\eta \end{pmatrix}, \qquad (9.45)$$

with angles β, γ, η defined as shown in

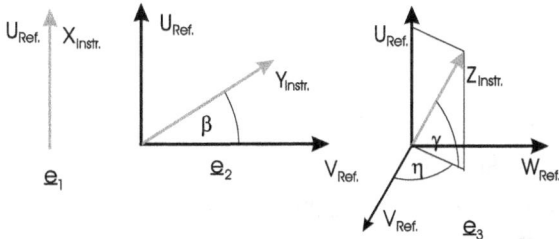

Figure 9.6: Geometrical interpretation of the matrix for orthogonalization to transform the co-ordinate system of the instrument into the reference co-ordinate system

The calculation of the angles of the three sensor axes within the facility co-ordinate system is given by:

$$\cos\xi_{xy} = \mathbf{e}_1 \cdot \mathbf{e}_2 = \begin{pmatrix} 1 \\ 0 \\ 0 \end{pmatrix} \cdot \begin{pmatrix} \sin\beta \\ \cos\beta \\ 0 \end{pmatrix}, \tag{9.46}$$

$$\cos\xi_{xz} = \mathbf{e}_1 \cdot \mathbf{e}_3 = \begin{pmatrix} 1 \\ 0 \\ 0 \end{pmatrix} \cdot \begin{pmatrix} \sin\gamma \\ \cos\gamma\sin\eta \\ \cos\gamma\cos\eta \end{pmatrix}, \tag{9.47}$$

$$\cos\xi_{yz} = \mathbf{e}_2 \cdot \mathbf{e}_3 = \begin{pmatrix} \sin\beta \\ \cos\beta \\ 0 \end{pmatrix} \cdot \begin{pmatrix} \sin\gamma \\ \cos\gamma\sin\eta \\ \cos\gamma\cos\eta \end{pmatrix}. \tag{9.48}$$

The multiplication gives:

$$\cos\xi_{xy} = \sin\beta, \tag{9.49}$$
$$\cos\xi_{xz} = \sin\gamma, \tag{9.50}$$
$$\cos\xi_{yz} = \sin\beta\sin\gamma + \cos\beta\cos\gamma\sin\eta. \tag{9.51}$$

A calculation exchanging all variables depending on the angles β, η, γ with those depending on ξ_{xy}, ξ_{xz} and ξ_{yz}

$$\underline{\underline{O}} = \begin{pmatrix} 1 & \cos\xi_{xy} & \cos\xi_{xz} \\ 0 & \sin\xi_{xy} & \frac{\cos\xi_{yz}-\cos\xi_{xy}\cos\xi_{xz}}{\sin\xi_{xy}} \\ 0 & 0 & \sqrt{1-\cos^2\xi_{xz}-\frac{(\cos\xi_{yz}-\cos\xi_{xy}\cos\xi_{xz})^2}{\sin^2\xi_{xy}}} \end{pmatrix}. \tag{9.52}$$

$\underline{\underline{D}}_{\text{Euler}}$ is calculated by separation of $\underline{\underline{E}}$ and $\underline{\underline{O}}$ of $\underline{\underline{M}}_2$. Comparing the coefficients of $\underline{\underline{D}}_{\text{Euler}}$ with the following formula given by an Euler rotation

$$\underline{\underline{D}}'_{\text{Euler}} = \begin{pmatrix} 1 & 0 & 0 \\ 0 & \cos\lambda & -\sin\lambda \\ 0 & \sin\lambda & \cos\lambda \end{pmatrix} \begin{pmatrix} \cos\mu & 0 & \sin\mu \\ 0 & 1 & 0 \\ -\sin\mu & 0 & \cos\mu \end{pmatrix} \begin{pmatrix} \cos v & -\sin v & 0 \\ \sin v & \cos v & 0 \\ 0 & 0 & 1 \end{pmatrix} \tag{9.53}$$

$$\underline{\underline{D}}'_{\text{Euler}} = \begin{pmatrix} \cos v\cos\mu & \sin v\cos\lambda + \cos v\sin\mu\sin\lambda & \sin v\sin\lambda - \cos v\sin\mu\cos\lambda \\ -\sin v\cos\mu & \cos v\cos\lambda - \sin v\sin\mu\sin\lambda & \cos v\sin\lambda + \sin v\sin\mu\cos\lambda \\ \sin\mu & \cos\mu\sin\lambda & \cos\mu\cos\lambda \end{pmatrix}$$

This matrix describes the rotation of the sensor triple against the reference co-ordinate system and allows to derive the angles of rotation.

9.6. MODELLING NON LINEAR EFFECTS

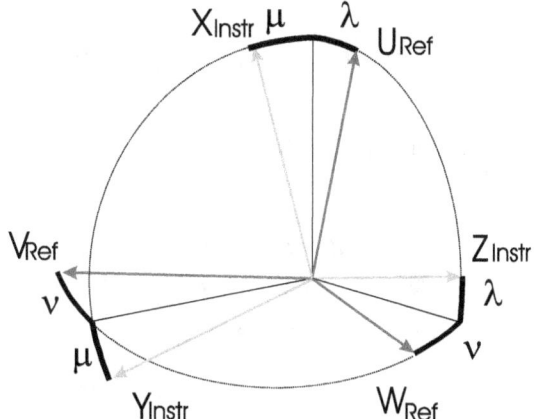

Figure 9.7: Geometric view on the orthogonalization of the instrument system and its transformation into the reference system

Out of a variety of possible Euler rotations the 3,2,1′ rotation is chosen, i.e. a rotation about the W–axis, then about the V–axis and finally about the U–axis. (In case of a rotation 1,3,1′ the matrix built up by partial rotations tends to become singular).

9.6 Modelling Non Linear Effects

9.6.1 Assumption of a Non Linear Sensitivity

The assumptions of a linear model can not be applied in case that the electronics have a non linear transfer function. This type of non linearity of electronics will be represented by a polynomial function within the model. e_1 from the linear model is replaced by

$$b_x = \sum_{i=0}^{N} e_i \, f_s^i \tag{9.54}$$

with

$$f_s = d_{11}b_u + d_{12}b_v + d_{13}b_w \tag{9.55}$$
$$= \cos(\angle[S,U])b_u + \cos(\angle[S,V])b_v + \cos(\angle[S,W])b_w. \tag{9.56}$$

Derived from Eq. (9.54) it is

$$b_x = \sum_{i=0}^{N} e_i \left(\cos(\angle[S,U])b_u + \cos(\angle[S,V])b_v + \cos(\angle[S,W])b_w \right)^i \tag{9.57}$$

with
$$1 = \cos(\angle[S,U])^2 + \cos(\angle[S,V])^2 + \cos(\angle[S,W])^2.$$

The number of relevant/required coefficients can be decreased by aligning the sensor during the calibration with the U–axis better than approximately $1°$.

$$0° \leq \angle[S,U] \leq 1° \qquad (9.58)$$
$$89° \leq \angle[S,V] \leq 90° \qquad (9.59)$$
$$89° \leq \angle[S,W] \leq 90° \qquad (9.60)$$

and therefore

$$1 \geq \cos(\angle[S,U]) = d_{11} > 0.9985 \qquad (9.61)$$
$$0.01745 \geq \cos(\angle[S,V]) = d_{12} > 0.0 \qquad (9.62)$$
$$0.01745 \geq \cos(\angle[S,W]) = d_{13} > 0.0. \qquad (9.63)$$

All coefficients where d_{12} or d_{13} are of higher order can be neglected in case that that the required/requested accuracy is less than the product of non linear coefficients e_i and the higher orders of the cosine of the angle $[S,V]$ and $[S,W]$. Then Eq.(9.57) will be

$$\begin{aligned} b_x =\ & e_0 \cos(\angle[S,U])b_u + e_0 \cos(\angle[S,V])b_v + e_0 \cos(\angle[S,W])b_w + \\ & + e_1 \cos(\angle[S,U])^2 b_u^2 + \cdots + e_N \cos(\angle[S,U])^N b_u^N. \end{aligned} \qquad (9.64)$$

Summarizing $e_0 \cos(\angle[S,U])b_u, \ldots$ to 'new' coefficients it is:

$$b_x = g_{10} b_u + g_{20} b_v + g_{30} b_w + \sum_{i=1}^{N} g_{1i} b_u^i \qquad (9.65)$$

The new equation system is linear in its coefficients and can therefore be solved using a least mean square fit of parameter.

9.6.2 Assumption of a Non Linear Dependency of Orientation

The model mentioned above is not applicable for the approximation of sensors influenced by the magnetic cross coupling effect. Therefore the model of a non linear dependency of orientation is introduced by

$$b_\mathsf{x} = \sum_{i=0}^{N} e_i \left(\sum_{j=0}^{M_1} a_{1j} \cos(\angle[S,U]) b_\mathsf{u}^j + \right.$$
$$+ \sum_{k=0}^{M_2} a_{2k} \cos(\angle[S,V]) b_\mathsf{v}^k +$$
$$\left. + \sum_{l=0}^{M_3} a_{3l} \cos(\angle[S,W]) b_\mathsf{w}^l \right)^i. \quad (9.66)$$

The coefficients a_{1j}, a_{2k}, a_{3l} introduce the non linear dependency of orientation. Again the coefficients $a_{1j} \cos(\angle[S,U])$,... are summarized by 'new' coefficients.

$$b_\mathsf{x} = \sum_{i=0}^{N} e_i \left(\sum_{j=0}^{M_1} g_{1j} b_\mathsf{u}^j + \sum_{k=0}^{M_2} g_{2k} b_\mathsf{v}^k + \sum_{l=0}^{M_3} g_{1l} b_\mathsf{w}^l \right)^i. \quad (9.67)$$

Typically it is assumed that the electronics of the magnetometer behaves linear during measurements regarding magnetic cross talking. Therefore it is set $N = 1$ and $e_1 = 1$. Then it is:

$$b_\mathsf{x} = \sum_{j=0}^{M_1} g_{1j} b_\mathsf{u}^j + \sum_{k=0}^{M_2} g_{2k} b_\mathsf{v}^k + \sum_{l=0}^{M_3} g_{1l} b_\mathsf{w}^l. \quad (9.68)$$

9.6.3 Model without any Pre-Knowledge of the Transfer Function

Setting $N > 1$ in Eq.(9.67) ordering the equation b_u, b_v and b_w and introducing new coefficients a_{ijk} leads to the following general model:

$$b_\mathsf{x} = \sum_{i=0}^{M} \sum_{j=0}^{N} \sum_{k=0}^{P} a_{ijk} b_\mathsf{u}^i b_\mathsf{v}^j b_\mathsf{w}^k. \quad (9.69)$$

This model allows to approximate up to 216 coefficients (maximal order of $M,N,P = 5$). This type of approximation is only one of the possibilities. Another possible approximation for example using spherical functions is described by [18].

9.7 Determination of Linear Model Coefficients

The calculation of the coefficients related to the linear model is done using the method of least mean square fit. The formulation of the basics of this

algorithm is described in [19, 10, 6, 5]. Especially [19] summarizes the techniques to find a numerical solution. This chapter mentions only a few of the relevant aspects without going too deeply into the mathematical details.
The method of least mean square fit is applied to the example in Equation(9.17)

$$b_{\text{Instr},x} = l_1 b_u + l_2 b_v + l_3 b_w.$$

The goal of this process is to minimize the differences between the model and the measurement of N calibration vectors.

$$\sum_{i=1}^{N} (l_1 \, b_{u,i} + l_2 \, b_{v,i} + l_3 \, b_{w,i} - b_{\text{Instr},i})^2 \stackrel{!}{=} \text{Min} \qquad (9.70)$$

One necessary condition to minimize the differences is that the first deviations of the coefficients l_1, l_2 and l_3 are zero.

$$\frac{\partial}{\partial l_1} \; : \; 2 \sum_{i=1}^{N} ((l_1 \, b_{u,i} + l_2 \, b_{v,i} + l_3 \, b_{w,i} - b_{\text{Instr},i}) \, b_{u,i}) = 0, \qquad (9.71)$$

$$\frac{\partial}{\partial l_2} \; : \; 2 \sum_{i=1}^{N} ((l_1 \, b_{u,i} + l_2 \, b_{v,i} + l_3 \, b_{w,i} - b_{\text{Instr},i}) \, b_{v,i}) = 0, \qquad (9.72)$$

$$\frac{\partial}{\partial l_3} \; : \; 2 \sum_{i=1}^{N} ((l_1 \, b_{u,i} + l_2 \, b_{v,i} + l_3 \, b_{w,i} - b_{\text{Instr},i}) \, b_{w,i}) = 0. \qquad (9.73)$$

This is a linear equation system in l_1, l_2 and l_3. After converting the terms and using

$$< \cdots > := \sum_{i=1}^{N} \cdots , \qquad (9.74)$$

$$\begin{pmatrix} < b_{u,i} \, b_{u,i} > & < b_{v,i} \, b_{u,i} > & < b_{w,i} \, b_{u,i} > \\ < b_{u,i} \, b_{v,i} > & < b_{v,i} \, b_{v,i} > & < b_{w,i} \, b_{v,i} > \\ < b_{u,i} \, b_{w,i} > & < b_{v,i} \, b_{w,i} > & < b_{w,i} \, b_{w,i} > \end{pmatrix} \begin{pmatrix} l_1 \\ l_2 \\ l_3 \end{pmatrix} = \begin{pmatrix} < b_{\text{Instr},i} \, b_{u,i} > \\ < b_{\text{Instr},i} \, b_{v,i} > \\ < b_{\text{Instr},i} \, b_{w,i} > \end{pmatrix}. \qquad (9.75)$$

The equation system (9.75) has a single solution as long as three independent calibration vectors are used.

9.8 Estimation of the Quality of Fit

The linear approximation will always lead to one single solution as long as there are at least three linear independent field configurations used for the approximation. But this allows no statement about the quality of the approximation in general. Using only three independent field configurations for approximation leads to a so called point approximation. This approximation fits the parameter of the model with respect to the measured value without any error. But it is not known how accurate the approximation would fit applying any other measured fields . It is possible to use so called test vectors to get a global statement of the quality of the fit. Each test vector is build up of a magnetic field configuration applied in the facility and the answer of instrument to it. Test vectors are not always used for the approximation of the coefficients. One part of the test vector either the magnetic field configuration or the measured value of the instrument is corrected using the coefficients. This part is called the 'Is' value. This value is compared to the part of the test vector the so called 'Shall' value:

$$\mathbf{b}_{Res} = \mathbf{b}_{Is} - \mathbf{b}_{Shall}. \qquad (9.76)$$

A statistical Analysis of the residual fields of many test vectors allows a statement regarding the quality of the fit. Minimum and maximum of residues show the maximum deviation found. Mean value and standard deviation show the expected residual field after correction of the data.

9.9 Offset of Magnetometer and Residual Field of Facility

Looking at the example from Section 9.3 it is obvious that the linear K-model of a magnetometer mentioned in Section 9.5 is not able to describe completely the measured values of the instrument. Especially a constant residual field different from zero is found after approximation for every sensor of the instrument. This residual field shown in Eq. (9.17)

$$b_{\text{Instr},x} = l_1 b_u + l_2 b_v + l_3 b_w$$

can be formally minimized by adding an additional constant coefficient l_0. Then it is

$$b_{\text{Instr},x} = l_0 + l_1 b_u + l_2 b_v + l_3 b_w. \qquad (9.77)$$

An offset b_{Offset} has to be applied to every component of the magnetometer. The offset value of a component is defined as the value seen in zero magnetic field [8, 16]. This offset value is typically linked to the residual magnetic field b_{ResFac} related to the facility. b_{ResFac} covers the magnetic field that is applicable in spite of the compensation of the facility. Assuming that there are no additional unknown magnetic field components we get:

$$l_0 = b_{\text{Offset}} + b_{\text{ResFac}}. \qquad (9.78)$$

9.9.1 Determination of Facility Offset and Residual Field

If the offset of the magnetometer and the residual field of the facility are combined it is not possible to separate both by just measuring in one geometrical orientation. Therefore the sensor is turned by 180 degree within the zero magnetic field of the facility. This separates offset and residual field of the facility. If b_1 is the first and b_2 the second 180 degree turn measurement inside of the zero field of the facility, then b_{ResFac} is found by

$$b_{\mathsf{ResFac}} = \frac{b_1 - b_2}{2}. \tag{9.79}$$

The offset b_{Off} can be calculated using the formula

$$b_{\mathsf{Off}} = \frac{b_1 + b_2}{2}. \tag{9.80}$$

9.9.2 Influence Elimination of Offset and Residual Field

Typically it is assumed that the inner parameter of the instrument are not variable with time. But the parameter of the residual field of the facility $\mathbf{b}_{\mathsf{ResFac}}$ is not always stable. Therefore it is often necessary to correct it to avoid a time based influence on the transfer function of the instrument.

Ideally $\mathbf{b}_{\mathsf{ResFac}}$ should be measured and corrected continuously. This requires a second sensor mounted close to the sensor to be calibrated. Due to the influence of this second sensor on the first one this is not feasible[4]. So called 'Zero' fields are measured at several moments at least at the beginning and the end of a calibration measurement to allow the detection of a possible variation of $\mathbf{b}_{\mathsf{ResFac}}$. 'Zero' field measurements are done by generating a magnetic zero field $\mathbf{b}_{\mathsf{Ref}}$ as exact as possible. This field is measured by the sensor. Looking at the time series of these measurements $\mathbf{b}_{\mathsf{Rest}}$ is varying during the measurement.

In case that the drifting of $\mathbf{b}_{\mathsf{Rest}}$ is slow with respect to the sample rate of the 'Zero' fields, it is possible to correct one component of a sensor $\underline{\mathbf{b}}_{\mathsf{Instr}}$ by subtracting the corresponding component of $\mathbf{b}_{\mathsf{Rest}}$ from it.

In the simplest case the vector $(\mathbf{b}_{\mathsf{Ref}}, \mathbf{b}_{\mathsf{Instr}})$ is determined within zero field and subtracted from all other following vectors. This allows to determine the influence of $\mathbf{b}_{\mathsf{ResFac}}$ and $\mathbf{b}_{\mathsf{Off}}$. The measurement of some 'Zero' fields at

---

[4] The additional second sensor is influencing the sensor to be calibrated if the distance between them is too close

the beginning, the end and in between calibration measurements helps to identify the variation of test set-up and or the instrument to be calibrated.

9.10 Transfer Function, Influence of Systematic Errors

The error after fitting of a test- or a calibration vector is partly given by systematic and statistical errors. The coefficients of a model are calculated based on the calibration vectors using a least mean square fit. The number of coefficients to be used for fitting follows on one hand the goal of minimizing the total sum of residuals. On the other hand a model is searched for minimizing the number of coefficients . This implies that the error of the mathematical model is direct depending on the error of each of its vectors used for calibration. The essential work during the process of analysis is found to be the detection and elimination of systematic errors in the data. In case that all systematic errors are eliminated only statistical errors will be observed. The statistical part of errors observed during physical measurements will follow the Gaussian deviation. On the other hand this implies that uncorrected systematic errors will have an impact if a distribution of errors other than the Gaussian is found.

A significant statement whether the statistics of the residual fields follow a Gaussian deviation or not can typically not be done if the number of vectors typically measured is too small. Therefore the residuals are plotted versus different calibration parameters (i.e. time, temperature, input values) for detecting of errors. In case that within this plot significant structures are found it is for sure that other systematic errors not covered by the mathematical model have an influence on the data. In this case it is necessary to do other investigations to identify the origin of the systematic effects.

9.11 Selection of Suitable Test Vectors

Section 9.7 shows that the selection of test vectors has an effect on the fitting. If no further pre information is available it is useful to base the fitting on a high even number of test vectors. The number of test vectors is limited by two factors. First of all it is necessary to run all measurements in a facility with a constant stable ambient field. Second the time available for measuring is typically limited. The following three sections describe different strategies to gain a reasonable amount of vectors for a fitting. Fig.9.8

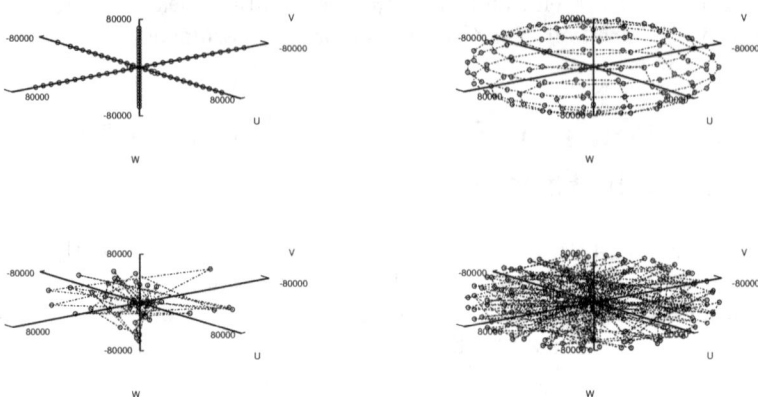

Upper left: Distribution of field configurations during an on–axis measurement; Upper right: Distribution of field configurations during a spike–sphere measurement on the surface of a sphere; Lower left: Distribution of field configurations during a spike–sphere measurement randomly with a enclosing sphere; Lower right: Distribution of field configurations during a spike–sphere measurement including the measurement of zero field every n^{th} vector;

9.11.1 On-Axis Measurements

This type of measurement the so called on-axis measurement was introduced in Section 9.3. This type of measurement is used since decades. The advantages of this type of measurement are found in the fact that magnetic fields of the facility are deviated in just one parameter at a time.

In case that the answer of the instrument is linear , geometrical straight lines will be found when plotting the measured values of the instrument versus applied fields of the facility. The gradient of these straight lines is correlated to one of the coefficients of the transfer matrix $\underline{\underline{M}}_1$ in Eq.(9.28). This allows the determination of the coefficients without fitting the parameters using the method of least mean square fit from Section 9.7 by applying a straight line to it. By this way all coefficients can be calculated directly. This point is essential because it allows a direct estimation of the result during the 'expensive' time in the facility. In addition it is possible to estimate the validity of each vector directly and decide to repeat the measurement step if

9.11. SELECTION OF SUITABLE TEST VECTORS

needed.

Some disadvantages are still related to this type of measurement. By using the gradient for determining the coefficients instead of a least mean square fit it is necessary that certain special magnetic field configurations are applied within the facility. If for example a magnetic field vector of the facility in direction of U shall be used for fitting then it is not possible to use a vector with small field values in V and W. This requires the ability of the facility to produce magnetic field configurations requested very accurate. Typically this is not the fact. Most often it is possible to produce a field configuration close to the requested one and to measure exactly its orientation and strength. For example a requested field configuration of $(45000, 0, 0)$ and $(50000, 0, 0)$ nT by the facility may produce a magnetic field configuration of $(45010, 15, 1)$ and $(50015, 8, -5)$ nT. These field configuration are not usable to calculate the coefficients using a straight line.

Even if the coefficients of the transfer function are calculated using a least mean square fit it is for sure that the used magnetic field configurations are only covering a small part along the axes of the magnetic space UVW (see Fig. 9.11 upper left). This coverage is sufficient for fitting a linear model. But it is not possible to detect more special non linearities of an instrument. Therefore an on-axis measurement can be used as a basic measurement especially to ensure the transfer function coefficients of a well known instrument and to determine magnetically the orientation of a linear instrument. In addition this type of measurement has to be extended when testing an unknown instrument.

9.11.2 Spike–Sphere–Measurement

The magnetic field configurations of a spike–sphere are build up in a way that the magnetic space is completely covered. This includes first a sequence of several field configurations on the surface of a magnetic sphere (demonstrated in Fig. 9.11 upper right), second a sequence of field configurations of random strength (smaller than the strength of sphere) and orientation (Fig. 9.11 lower left). Both sequences combined with a measurement of the zero magnetic field every n^{th} vector are building the complete spike–sphere (Fig. 9.11 lower right). The transfer function can be easily determined by using a least mean square fit.

References from chapter 9

1. Acuña M.H., Scearce C.S., Seek J.B., Scheifle J.: The MAGSAT Vector Magnetometer – A precision Fluxgate Magnetometer for the Measurement of the Geomagnetic Field; NASA Technical Memorandum 79656, 1978.

2. Acuña M.H.: MAGSAT – Vector Magnetometer Absolute Sensor Alignment Determination; NASA Technical Memorandum 79648, 1981.

3. Afanassiev J. W.: Ferrosonden–Geräte; Übersetzung aus dem Russischen; Institut für Geophysik und Meteorologie, Technische Universität Braunschweig, 1994.

4. Brauer P.: The Ringcore Fluxgate Sensor. Department of Automation, The Technical University of Denmark, June 1997

5. Böhm W., Gose G., Kahmann J.: Methoden der numerischen Mathematik. Wiesbaden: Vieweg, 1985.

6. Bronstein, Semendjajew: Taschenbuch der Mathematik. Neubearbeitung; BSB B. G Teubner Verlagsgesellschaft, Leipzig, und Verlag Nauka, Moskau, 1979.

7. Huber P.J.: Robust Regression – Asymtotics, Conjectures and Monte Carlo. Anals. of Statistics 1 (1973), pp. 799-821.

8. Kertz W.: Einführung in die Geophysik I u. II. 1. Aufl.; Mannheim: Bibliographisches Institut,1969

9. Korth H.: Untersuchung der physikalischen Eigenschaften eines Vektormagnetometers mit Netzbandkern. Diplomarbeit. Institut für Geophysik und Meteorologie der Technischen Universität Braunschweig, 1998.

10. Kowalsky H.J.: Lineare Algebra. 9. Überarb. u. erw. Aufl. -Berlin, New York: de Gruyter, 1979.

11. Kügler H.: Modell eines Saturationskernmagnetometers basierend auf hochgenauen Kalibriermessungen, Dissertation, URL:http://www.digibib.tu-bs.de /?docid=00001591, 2004.

9.11. SELECTION OF SUITABLE TEST VECTORS

12. Kuhnke F., Menvielle F., Musmann G., Karczewski J.F., Kügler H., Cavoit C., Schibler P.: The OPTIMISM/MAG Mars-96 experiment: magnetic measurements onboard landers and related magnetic cleanliness program. Planet. Space Sci. Vol. 46, No. 6/7, PP. 749-767,1998.

13. Kupke W.: Verkleinerung der Nichtlinearität und des Temperaturkoeffizienten eines Fluxgatemagnetometers mit Ringkernsensor. Diplomarbeit. Institut für Geophysik und Meteorologie der Technischen Universität Braunschweig, 1984.

14. Lühr H.: Das Impulsintegrationsverfahren. Dissertation. Geophysik und Meteorologie der Technischen Universität Braunschweig, 1980.

15. Lühr H.: MAGNETSRODE; Eine Einrichtung für magnetische Kalibrierungen und Testmessungen. GAMMA 42,Institut für Geophysik und Meteorologie, Technische Universität Braunschweig, 1984.

16. Ness N.F.: Magnetometers for Space Research; Space Sience Reviews **11**(1970) 459–554. Vetterding W.T., Flannery B.P.: Numerical Recipes in C. Cambridge, 2nd Edition 1994.

17. Primdahl F.: The fluxgate magnetometer. J.Phys.E, Scient.Inst.12,241-353. 1979.

18. Tarantola A.: Methods for Data Fitting and Model Parameter Estimation. Elsevier, 1987.

19. Überhuber C.: Computer Numerik1 und 2. 1. Aufl; Berlin Heidelberg:Springer,1995.

Chapter 10

Magnetometer Inflight Calibration

R.A.Langel[1]†, T.J.Sabaka[2],

[1] formerly of Goddard Space Flight Center,
Greenbelt, Md., USA
[2] Hughes/STX Corp., Lanham, Md.USA

Abstract:

Once in orbit, magnetometers onboard spacecraft, and their attitude relative to a known spacecraft coordinate system, are not subject to direct calibration. Although modern fluxgate magnetometers can be very reliable and stable, apart from comparison with an absolute standard, one can never be certain that drift has not occurred. Similarly, one cannot be certain that the vector magnetometer has not moved from its pre-launch position due to vibration or temperature effects. Methods for inflight calibration are needed. Such methods can be incorporated into the normal flow of data processing. Calibration of the fluxgate instrument parameters is accomplished by comparison with an absolute scalar instrument located near the fluxgate on the spacecraft. This calibration is independent of knowledge of spacecraft position, magnetometer attitude, or amount of magnetic disturbance. It does presume that the calibration parameters have not changed drastically from an accurate pre-flight calibration, and that the calibration parameters do not vary significantly in the time required to acquire enough data for the calibration procedure, several orbits. Assuming that the spacecraft is earth-oriented, any inaccuracy in the knowledge of that orientation will result in field perturbations that vary systematically with spacecraft orientation rather than with position in the earth-fixed coordinate system. This permits inclusion of the parameters of the transformation matrix between magnetometer and spacecraft coordinates as part of the procedure for determining a spherical harmonic model of the main field. Such a solution again depends upon the stability of the transformation, i.e., that it does not change significantly over the day or so required to acquire the data needed for deriving the main field model.

10.1 Non-Absolute Satellite Vector Magnetic Field Data

10.1.1 Introduction

Near-Earth magnetic field measurements suitable for study of the fields originating in the core and lithosphere must be of 1–5 nT accuracy. Prior to the Magsat mission, this was achieved only for measurements of the scalar, or total field. The Magsat mission successfully measured the vector components to an accuracy mostly better than 5 nT. Missions are now in preparation and others being planned to duplicate or improve upon the data quality achieved by Magsat. These include the Danish Ørsted satellite, the Argentinian SAC-C satellite, and the German CHAMP satellite. In order to achieve these accuracies, not only must the instrumentation used be of the highest accuracy, but the data must be carefully processed. Two problems that should be addressed in the data processing are adjustment of the calibration of the vector instrument and the verification and, if necessary, adjustment of the attitude, or pointing direction, of the vector data. One approach to addressing these problems is presented in this paper.

Why must the vector data calibration be adjusted? In general, vector magnetometers, usually of the fluxgate type, are not absolute but are subject to variations in electronic components that can lead to changes in calibration parameters. Modern fluxgates, when constructed with care, are very reliable, generally with little or no drift. However, apart from comparison with an absolute standard, one can never be certain that drift has not occurred. This leads to the notion of including an absolute scalar instrument on the spacecraft, located such as to measure the same field as the vector instrument to better than 1 nT, and of adjusting the vector instrument calibration parameters using the absolute measurements as a standard.

Why must the attitude solution be verified or adjusted? In order to achieve 5 nT accuracy in a magnetic component measurement in a 50 000 nT ambient field, the direction of that measurement must be known to within

20 arc seconds. This is achievable with modern star sensors. Furthermore, it is now possible to make star sensors that generate very low magnetic fields so they can be located on the same platform, or optical bench, as the vector magnetometer sensor. It is then necessary to accurately determine the rotation matrix between the magnetometer axes and the coordinate system of the star sensor. However, any mechanical shift after such determination is made, eg., during launch or because of thermal variation in orbit, can modify that relationship. Hence some method of verifying and/or adjusting the attitude information is needed.

It is useful to place the data processing used to address these problems in the context of a hypothetical overall scheme of the flow of data processing. Such a scheme is illustrated in Fig. 10.1. In this scheme, Level 0 data are the raw data from the spacecraft, Level 1 data has had calibration information and, possibly, coordinate adjustments applied, Level 2 data has undergone special selection, and Level 3 are interpretational products. The experiment data comprises magnetometer and star camera output. Ephemeris information is shown entering the data flow from a separate source. This scheme is very similar to that used in the Magsat program.

10.1. NON-ABSOLUTE SATELLITE VECTOR MAGNETIC FIELD DATA

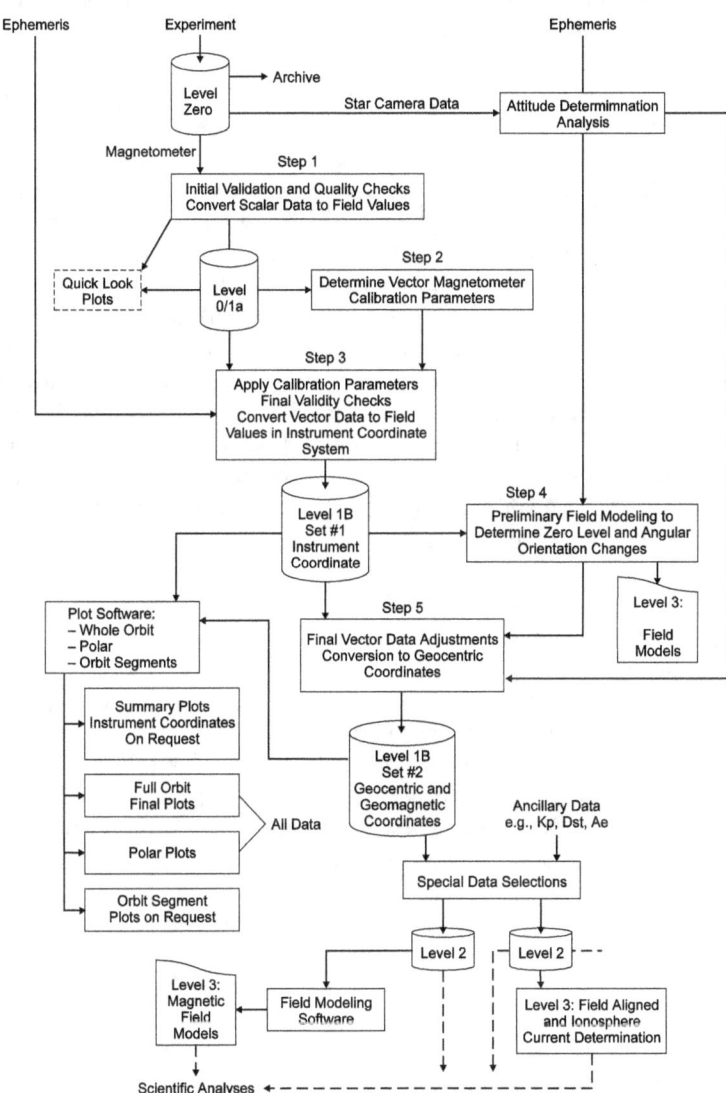

Figure 10.1: Diagram showing proposed data flow for a satellite magnetic field experiment.

At the start of processing, the star tracker data are split off and processed separately to determine the attitude of the magnetometer platform. The magnetometer data itself are subject to several steps of processing. The first step consists of initial validation and quality checks, eg. discarding data with instrument power off, verifying the time tag quality, checking instrument housekeeping, etc. In the second step a subset of the data is selected, both scalar and vector, and the scalar data are used to determine the calibration parameters of the vector data. Suggested procedures for this step are a main topic of this paper. They require approximately one day of data and are performed at one or two week intervals. The actual interval must be determined post launch and depends upon the stability of the calibration parameters. Step three is the application of the calibration parameters to the vector data and conversion of that data to magnetic field values. The three components are still in the instrument coordinate system. Additional validity checks may be performed at this time, eg. comparison of the field magnitude computed from the vector data with that measured by the scalar magnetometer. In step four, preliminary models of the Earth's field are derived. This step is also addressed in this paper because parameters solved for in these models may include any scale factor, zero level and/or angular orientation changes needed in the vector field measurements. The model outputs constitute a Level 3 output for use by investigators. Step five applies any changes in the zero level and/or angular orientation to the vector data and uses the attitude determination to rotate the vector data to a standard (r, θ, ϕ) Earth fixed geocentric coordinate system. At this step, geomagnetic coordinates are computed and added to the data set. Step six is the selection of any specialized data sets, eg. by magnetic disturbance level, by geographic position, by decimation level. It is anticipated that steps one through four will be performed only one time, unless some unforeseen problem arises.

In the following sections recommended procedures are presented for intercomparison of data from an absolute and a vector magnetometer and for using main field modeling procedures for determining angular orientation changes

10.1.2 Calibration of a Vector Magnetometer with Absolute Scalar Data

Procedures for this calibration have been documented for the Magsat mission by Lancaster et al. (1980) and Langel et al. (1981). A somewhat more generalized procedure has been documented by Langel et al. (1997) and will be followed here.

10.1. NON-ABSOLUTE SATELLITE VECTOR MAGNETIC FIELD DATA

Basic Formalism

The notation used is:

N	is the number of measurements used for the adjustment estimation,
$\mathbf{v}_i : i = 1, N$	are the measured magnetometer voltages,
$(\mathbf{B}_m)_i : i = 1, N$	are the actual field values along the actual magnetometer axes,
$(\mathbf{B}_0)_i : i = 1, N$	are the corresponding magnetic field values along orthogonal magnetometer axes,
$B_i : i = 1, N$	are the corresponding set of scalar field values from an absolute instrument,
$\mathbf{b}$	is any offset in the vector magnetometer,
μ_{12}	is the cosine of the angle between 1st and 2nd magnetometer sensor axes,
μ_{23}	is the cosine of the angle between 2nd and 3rd magnetometer sensor axes,
μ_{13}	is the cosine of the angle between 1st and 3rd magnetometer sensor axes,

where a little arrow on top indicates a vector quantity. Then define

$$\gamma_{ij} = \cos^{-1}(\mu_{ij}), \qquad \gamma'_{ij} = \gamma_{ij} - \frac{\pi}{2}. \qquad (10.1)$$

The satellite may be stabilized by magnetic torquers that result in a magnetic field at the magnetometers. This can be taken into account provided the torquer currents, $\mathbf{i}_t$, are measured. If $\mathbf{B}_t$, is the field from the torquer bars at the magnetometer, then for the ith measurement, the adjustment model is

$$(\mathbf{B}_m)_i = S\mathbf{v}_i + \mathbf{b} + T(\mathbf{i}_t)_i \qquad (10.2)$$
$$(\mathbf{B}_0)_i = H(\mathbf{B}_m)_i, \qquad (10.3)$$

where S is the matrix of instrument scale factors, T is a matrix relating torquer current to the field along the magnetometer axes, and H is the transformation matrix to the orthogonal magnetometer coordinate system. For simplicity, S is assumed diagonal. However the results can readily be extended for the non-diagonal case. Here, the diagonal elements, $\{S_{jj}, j = 1, 3\}$ are to be determined. H is not diagonal, but its elements depend only upon the three μ_{ij}, or γ'_{ij}, defined above. If

$$\chi = \sum_i [|(\mathbf{B}_0)_i| - B_i]^2, \qquad (10.4)$$

222   CHAPTER 10. MAGNETOMETER INFLIGHT CALIBRATION

then the procedure is to linearize (10.4) with respect to the variables in the solution and find the least squares minimum. Because (10.4) is non-linear in the variables, an iterative approach is required, eg., if superscript k indicates the variable value after the kth iteration then set:

$$\mathbf{b}^{k+1} = \mathbf{b}^k + \delta\mathbf{b}^k \,, \tag{10.5}$$
$$S^{k+1} = S^k + \delta S^k \,, \tag{10.6}$$
$$H^{k+1} = H^k + \delta H^k \,, \tag{10.7}$$
$$T^{k+1} = T^k + \delta T^k \,. \tag{10.8}$$

The quantity χ is then minimized with respect to the perturbation variables, $\delta\mathbf{b}^k$, δS^k, δH^k, and δT^k.

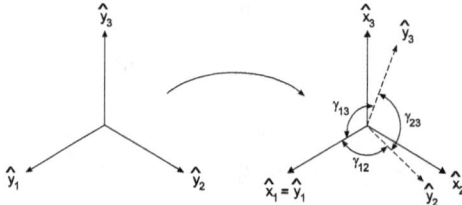

Figure 10.2: Orthogonal, $\hat{\mathbf{x}}$, and non-orthogonal, $\hat{\mathbf{y}}$, magnetometer axes, and the associated angles.

Definitions of $\mathbf{b}^k$, $\delta\mathbf{b}^k$, $(\gamma'_{ij})^k$, $(\delta\gamma'_{ij})^k$, S^k, δS^k, T^k, and δT^k are straightforward. The matrix H transforms the vector $\mathbf{B}_m$, with unit vectors in the non-orthogonal directions $\hat{\mathbf{y}}_i$, $i = 1,2,3$, to an orthogonal system with unit vectors $\hat{\mathbf{x}}_i$, $i = 1,2,3$, defined as follows, see Fig. 10.2. The $\hat{\mathbf{x}}_1$ axis is taken to be equal to the $\hat{\mathbf{y}}_1$ axis; the $\hat{\mathbf{x}}_2$ axis is taken to be in the plane formed by the $\hat{\mathbf{y}}_1$ and $\hat{\mathbf{y}}_2$ axes and oriented as near as possible to the $\hat{\mathbf{y}}_2$ axis; and the $\hat{\mathbf{x}}_3$ axis is taken such as to form a right handed system. $\hat{\mathbf{x}}_3$ will be nearly along the $\hat{\underline{\mathbf{y}}}_3$ axis because the $\hat{\mathbf{y}}$ axes are themselves nearly orthogonal.

By these definitions

$$\hat{\mathbf{y}}_2 \cdot \hat{\mathbf{x}}_3 = 0, \quad \hat{\mathbf{y}}_i \cdot \hat{\mathbf{y}}_i = \hat{\mathbf{x}}_i \cdot \hat{\mathbf{x}}_i = 1 \,. \tag{10.9}$$

Noting that

$$x_i = x_1\hat{\mathbf{x}}_1 + x_2\hat{\mathbf{x}}_2 + x_3\hat{\mathbf{x}}_3 = y_1\hat{\mathbf{y}}_1 + y_2\hat{\mathbf{y}}_2 + y_3\hat{\mathbf{y}}_3 \,, \tag{10.10}$$

and using (10.9), it can be shown that

$$H_{11} = 1.0, \quad H_{12} = \mu_{12}, \quad H_{13} = \mu_{13}, \qquad (10.11)$$
$$H_{21} = H_{31} = H_{32} = 0.0, \qquad (10.12)$$
$$H_{22} = \left[1 - \mu_{12}^2\right]^{\frac{1}{2}}, \qquad (10.13)$$
$$H_{23} = \left(\frac{\mu_{23} - \mu_{12}\mu_{13}}{(1 - \mu_{12}^2)^{\frac{1}{2}}}\right), \qquad (10.14)$$
$$H_{33} = \left(1 - \mu_{13}^2 - \frac{(\mu_{23} - \mu_{12}\mu_{13})^2}{1 - \mu_{12}^2}\right)^{\frac{1}{2}}. \qquad (10.15)$$

In practice it is easier to work with the γ'_{ij}. In particular, if the γ'_{ij} are very small

$$\mu_{ij} = \cos\left(\gamma'_{ij} + \frac{\pi}{2}\right) = -\sin\gamma'_{ij} \doteq \gamma'_{ij}, \qquad (10.16)$$
$$\left[1 - \mu_{ij}^2\right]^{\frac{1}{2}} = \sin\gamma_{ij} = \cos\gamma'_{ij} \doteq 1 - \frac{(\gamma'_{ij})^2}{2}, \qquad (10.17)$$
$$1 - \mu_{ij}^2 \doteq 1 - (\gamma'_{ij})^2, \qquad (10.18)$$

so that

$$H(\gamma') = \begin{bmatrix} 1 & -\gamma'_{12} & -\gamma'_{13} \\ 0 & 1 - \frac{\gamma'^2_{12}}{2} & \frac{-\gamma'_{23} - \gamma'_{12}\gamma'_{13}}{1 - \gamma'^2_{12}/2} \\ 0 & 0 & \sqrt{1 - \gamma'^2_{13} + \frac{(\gamma'_{23} + \gamma'_{12}\gamma'_{13})^2}{1 - \gamma'^2_{12}}} \end{bmatrix} \qquad (10.19)$$

Expressions for the δH^k are given in Langel et al (1997).

Extended Formalism

The model of Langel et al. (1981) assumes that the fluxgate response is linear in the voltages, **v**. This is not always the case and it may be necessary to add terms proportional to higher powers of the components, $\mathbf{v}_i$. In some instruments the range of the fluxgate sensors are limited, eg., to ± 1000 to ± 2500 nT, as on the Magsat spacecraft. Extension of the total instrument range to the $\pm 64\,000$ nT needed to measure the Earths field is then accomplished by injecting precisely known currents into coils surrounding each sensor, resulting in uniform bias fields along each sensor axis. Current strength is varied stepwise so that the bias magnetic field from the coil cancels most of the ambient field, allowing the sensor to function in the desired range. The sum of the bias field and the field measured by the fluxgate sensor then equals the ambient field along the sensor axis. For Magsat, seven current steps were employed for each axis. Extension of (10.2) can then be accomplished by adding a term of the form $A\omega$, where A is a matrix of dimension 3 by the number of current steps, 3 by 21 for Magsat since there are 7 current steps for each magnetometer axis, with entries equal to 0 if that current step is not activated and equal to 1 if it is activated, and where ω is a vector with entries equal to the field values produced by the corresponding current step. For Magsat, ω is of dimension 21. Note that, for Magsat, at most seven entries in each row of A are non zero. The ω_i are then among the parameters determined during the in-flight calibration.

A fluxgate instrument may also exhibit dependence on temperature, both of the sensor and of the electronics. Let τ_s and τ_e be the deviation of the sensor and electronic temperatures from their nomimal values. Then (10.2) is extended by addition of the terms $\tau_s \mathbf{t}_s + \tau_e \mathbf{t}_e$, where $\mathbf{t}_s$ and $\mathbf{t}_e$ are vectors of temperature scale factors for each axis of the magnetometer, to be solved for in the calibration computation.

The magnetometer used for Magsat also exhibited what is called crosstalk between the three magnetometer axes. That is, the measured field along each axis showed a small dependence upon the field along the other two axes. Terms to account for this effect are described in Lancaster et al. (1980).

Combining terms, excluding nonlinear and cross-talk effects, (10.2) is expanded to become

$$(\mathbf{B}_m)_i = S\mathbf{v}_i + \mathbf{b} + T(\mathbf{i}_t)_i + (A)_i \omega + (\tau_s)_i \mathbf{t}_s + (\tau_e)_i \mathbf{t}_e , \qquad (10.20)$$

which is then the general equation for in-flight calibration.

The Ørsted fluxgate magnetometer is a triaxial fluxgate inside a spherical coil system. The coil system is used to null the field at the fluxgate sensors. This means that the measurement, $\mathbf{v}$, comprises the voltages across three precision resistors, one for each coil axis. The stability of the measurement depends upon the mechanical stability of the spherical coil system and the stability of the electronic components. Equation (10.20), without the term in $A\omega$, is the appropriate calibration equation.

Iterative Solution with a priori

Now let $\mathbf{p}$ be the vector of calibration parameters with $\mathbf{p}^0$ an a priori starting estimate with covariance matrix Ω_0 on $\mathbf{p} - \mathbf{p}_0$, possibly resulting from a pre-flight calibration. What is sought is an estimate of $\mathbf{p}$ that minimizes the differences between the $|(\mathbf{B}_0)_i|$, and the B_i, in a least squares sense, ie., χ.

10.1. NON-ABSOLUTE SATELLITE VECTOR MAGNETIC FIELD DATA

Suppose that after the kth iteration the solution is $(\mathbf{B}_0)_i^k$ and that the next iteration results in the solution

$$(\mathbf{B}_0)_i^{k+1} = (\mathbf{B}_0)_i^k + \delta(\mathbf{B}_0)_i^k . \tag{10.21}$$

Now the scalar difference after the kth iteration is given by

$$\delta B_i^k = B_i - |(\mathbf{B}_0)_i^k| \doteq \frac{\delta(\mathbf{B}_0)_i^k \cdot (\mathbf{B}_0)_i^k}{|(\mathbf{B}_0)_i^k|} . \tag{10.22}$$

The approximation in (10.22) defines the relationship between the scalar difference after the kth iteration, δB_i^k, and the quantity $\delta(\mathbf{B}_0)_i^k$ which is to be estimated in the $k+1$ iteration. If the estimate of $\underline{\mathbf{p}}$ after the kth iteration is $\mathbf{p}^k$, then the $(k+1)$ iteration, $\mathbf{p}^{k+1}$, is given by

$$\mathbf{p}^{k+1} = \mathbf{p}^k + \delta \mathbf{p}^k . \tag{10.23}$$

Equation (10.22) comprises part of the problem linearization. It is also necessary to linearize (10.3), using (10.20). Here it is assumed that the terms in $A\omega$ are not required, e.g.,

$$\begin{aligned}
(\mathbf{B}_0)_i^{k+1} &= (H^k + \delta H^k)\left[(S^k + \delta S^k)\mathbf{v}_i + \mathbf{b}^k + \delta \mathbf{b}^k + (T^k + \delta T^k)(\mathbf{i}_t)_i \right. \\
&\quad \left. + (\tau_s)_i(\mathbf{t}_s^{\ k} + \delta \mathbf{t}_s^{\ k}) + (\tau_e)_i(\mathbf{t}_e^{\ k} + \delta \mathbf{t}_e^{\ k})\right] \\
&= (\mathbf{B}_0)_i^k + H^k\left[\delta S^k \mathbf{v}_i + \delta \mathbf{b}^k + \delta T^k(\mathbf{i}_t)_i + (\tau_s)_i \delta \mathbf{t}_s^{\ k} + (\tau_e)_i \delta \mathbf{t}_e^{\ k}\right] \\
&\quad + \delta H^k \left[S^k \mathbf{v}_i + \mathbf{b}^k + T^k(\mathbf{i}_t)_i + (\tau_s)_i \mathbf{t}_s^{\ k} + (\tau_e)_i \mathbf{t}_e^{\ k}\right] \\
&\quad + \text{higher order terms.}
\end{aligned} \tag{10.25}$$

Omitting the higher order terms constitutes the required linearization giving

$$\begin{aligned}
\delta(\mathbf{B}_0)_i^k &= (\mathbf{B}_0)_i^{k+1} - (\mathbf{B}_0)_i^k \tag{10.26} \\
&= H^k\left[\delta S^k \mathbf{v}_i + \delta \mathbf{b}^k + \delta T^k(\mathbf{i}_t)_i + (\tau_s)_i \delta \mathbf{t}_s^{\ k} + (\tau_e)_i \delta \mathbf{t}_e^{\ k}\right] \tag{10.27} \\
&\quad + \delta H^k \left[S^k \mathbf{v}_i + \mathbf{b}^k + T^k(\mathbf{i}_t)_i + (\tau_s)_i \mathbf{t}_s^{\ k} + (\tau_e)_i \mathbf{t}_e^{\ k}\right] .
\end{aligned}$$

The collection $\{\delta S^k, \delta \mathbf{b}^k, \delta T^k, \delta \mathbf{t}_s^{\ k}, \delta \mathbf{t}_e^{\ k} \text{ and } \delta \gamma^k \text{ (i.e., elements of } \delta H^k)\}$ comprise $\delta \mathbf{p}^k$. If, then, $\delta \mathbf{c}^k$ is the N dimension vector of δB_i^k at the $i = 1, \ldots N$ locations, then substitution of (10.27) into (10.22) gives a collection of N equations that can be collectively written in the form

$$\delta \mathbf{c}^k = D^k \delta \mathbf{p}^k . \tag{10.28}$$

Equation (10.28) is overdetermined and it is natural to seek a least squares solution with parameter estimation given by

$$\delta \mathbf{p}^k = \left[(D^k)^\mathrm{T} W(D^k) + \lambda \Omega_0^{-1} \right]^{-1} \left[(D^k)^\mathrm{T} W \delta \mathbf{c}^k + \lambda \Omega_0^{-1} (\mathbf{p}^0 - \mathbf{p}^k) \right], \quad (10.29)$$

where W is a suitably chosen weight matrix and λ is a damping factor controlling the relative weight of the em a priori estimate, $\mathbf{p}^0$. The value of λ determines a tradeoff between how close the solution is to the a priori, $\mathbf{p}^0$, and the goodness of fit to the data, the χ of (10.4). For smaller λ, the solution will be dominated by the data, ie., $(D^k)^\mathrm{T} W(D^k)$ will dominate over $\lambda \Omega^{-1}$, and χ will be minimized. As λ increases, the solution will tend toward $\mathbf{p}^0$.

In-flight data may not be sufficient to determine all of the calibration parameters. For example, one axis of the Magsat magnetometer was oriented approximately in the east-west direction. As a result, the field magnitudes measured along that axis were generally less than 12 000 nT in magnitude, far below the instrument range. In this case, for Magsat, only combinations of parameters could be determined (Lancaster et al., 1980). Where all of the calibration parameters are not resolved by the data, the solution may be unstable, exhibiting undesirable behavior for some measurement directions and magnitudes. Stability is furnished by appropriate choice of $\mathbf{p}^0$ and λ, and can be evaluated by examination of the solution covariance matrix,

$$Cov\left(\mathbf{p} - \mathbf{p}^{k+1}\right) = \left[(D^k)^\mathrm{T} W(D^k) + \lambda \Omega_0^{-1} \right]^{-1}, \quad (10.30)$$

and by examination of the tradeoff curve between λ and χ. Typically, as λ increases from a low value, χ will change very little or increase slowly. At some point as λ increases, χ may begin to grow, ie., there will be a knee to the curve. Below that point, instability may occur; above that point the solution is usually stable. If a well defined knee to the curve is present, it generally indicates the optimum choice for λ. An alternative approach that might be appropriate if there is reason to think that the instrument calibration has changed significantly so that it is no longer close to $\mathbf{p}^0$ is to set $\mathbf{p}^0 = 0$ and replace $\lambda \Omega_0^{-1}$ by λI, where I is the identity matrix. This is the formalism for ridge regression. In this case the tradeoff is between solution smoothness, ie., $\mathbf{p} \cdot \mathbf{p}$, and goodness of fit to the data, χ. Examination of the covariance and tradeoff curve should be an integral part of any calibration procedure.

Formally, it is not necessary to perform the linearization of (10.27). From (10.2) and (10.3), suppressing the term for torquer current,

$$(\mathbf{B}_0)_i = H \left[S \mathbf{v}_i + \mathbf{b} + \ldots \right]. \quad (10.31)$$

10.1. NON-ABSOLUTE SATELLITE VECTOR MAGNETIC FIELD DATA

Then the elements of D^k are found from

$$D_{ij} = \frac{\partial (\mathbf{B}_0)_i}{\partial p_j} ,\qquad (10.32)$$

eg.,

$$\left.\frac{\partial (\mathbf{B}_0)_i}{\partial \gamma'_{pq}}\right|_k = \left.\frac{\partial H}{\partial \gamma'_{pq}}\right|_k \left(S^k \mathbf{v}_i + \mathbf{b}^k + \ldots \right) ,\qquad (10.33)$$

and

$$\left.\frac{\partial (\mathbf{B}_0)_i}{\partial S_{pp}}\right|_k = H^k \left(\left.\frac{\partial S}{\partial S_{pp}}\right|_k \mathbf{v}_i + \mathbf{b}^k + \ldots \right) ,\qquad (10.34)$$

with $|_k$ indicating that the quantity is evaluated using the values determined after the kth iteration.

10.1.3 Calibration of Magnetometer Attitude

A spacecraft platform is moving, and rotating, relative to the main field of the Earth. Thus a bias in the attitude knowledge, which is fixed relative to the spacecraft, will cause non-constant field changes in the earth-fixed system, as illustrated in Fig. 10.3. For example, an attitude bias in "roll" will result in a sawtooth wave in the earth-fixed Y (east) component. Thus, parameters describing the orientation of the magnetometer are resolvable in the data. This makes possible incorporation of those parameters into the formalism of main field modeling, as indicated in Step 4 of Fig.10.1.

Basic Formalism

Additional notation used is:
$\{(\mathbf{B}_G)_i: i = 1, N\}$ are field values in $\mathbf{r} = (r, \theta, \phi)$, earth fixed coordinates,
$\{(\mathbf{B}_{S/C})_i: i = 1, N\}$ are field values in spacecraft coordinates, i.e., the coordinates defined by the star sensors,
T_{GS} is the transformation matrix between spacecraft and earth-fixed coordinates,
$T_{S/C}$ is the transformation matrix from orthogonal magnetometer axes to the spacecraft coordinate system.

Then

$$\mathbf{B}_G = T_{GS} \mathbf{B}_{S/C} = T_{GS} T_{S/C} \mathbf{B}_0 .\qquad (10.35)$$

T_{GS} is a function of spacecraft position, $\mathbf{r}$, and spacecraft attitude is assumed to be known. $T_{S/C}$ is formulated as a function of the Euler angles, ε, defining the transformation from the orthogonal magnetometer axes to the spacecraft, or star sensor, coordinate axes. The definition of the orientation parameters $\varepsilon_1, \varepsilon_2, \varepsilon_3$ is shown in Fig. 10.4 with the order of rotation:

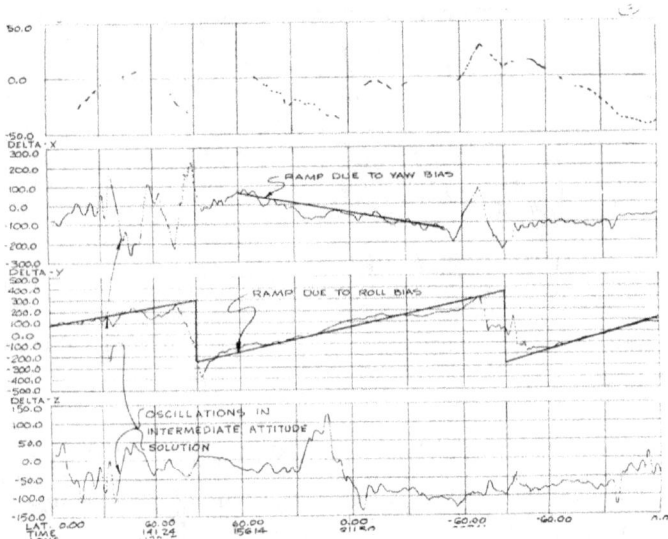

Figure 10.3: Residual data from the Magsat spacecraft illustrating the effects
of errors in the knowledge of magnetometer axes. The data shown are reduced
relative to a preliminary (intermediate) estimate of magnetometer attitude. The effects of biases, or errors, in the spacecraft yaw and roll axes are indicated.

1. rotate by ε_1 about the y-axis,

2. rotate by ε_2 about the new z-axis,

3. rotate by ε_3 about the new x-axis.

The resulting transformation matrix is:

$$T_{\text{S/C}} = \begin{pmatrix} \cos\varepsilon_1 \cos\varepsilon_2 & -\sin\varepsilon_2 & -\sin\varepsilon_1 \cos\varepsilon_2 \\ \sin\varepsilon_1 \sin\varepsilon_3 + \cos\varepsilon_3 \sin\varepsilon_2 \cos\varepsilon_1 & \cos\varepsilon_3 \cos\varepsilon_2 & \cos\varepsilon_1 \sin\varepsilon_3 - \sin\varepsilon_1 \sin\varepsilon_2 \cos\varepsilon_3 \\ \sin\varepsilon_1 \cos\varepsilon_3 - \cos\varepsilon_1 \sin\varepsilon_2 \sin\varepsilon_3 & -\sin\varepsilon_3 \cos\varepsilon_2 & \cos\varepsilon_3 \cos\varepsilon_1 + \sin\varepsilon_1 \sin\varepsilon_2 \sin\varepsilon_3 \end{pmatrix}$$
(10.36)

If the $(\mathbf{B}_0)_i$ are the measurements used in the main field modeling procedure, then

$$(\mathbf{B}_0)_i = T_{\text{S/C}}^{\text{T}} (T_{\text{GS}}^T)_i (\mathbf{B}_{\text{G}})_i \,, \qquad i = 1,\ldots,N \,, \tag{10.37}$$

10.1. NON-ABSOLUTE SATELLITE VECTOR MAGNETIC FIELD DATA

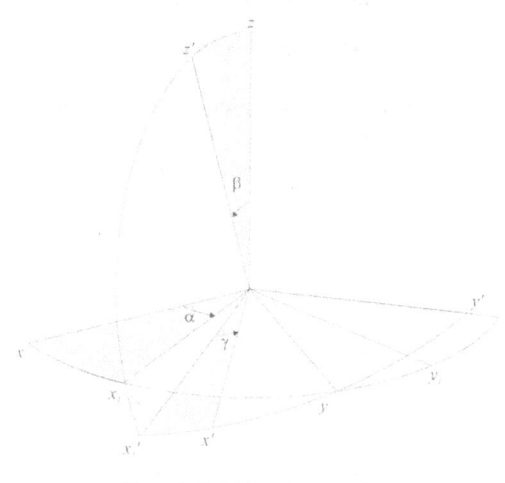

Figure 10.4: Euler angles of rotation between the orthogonal magnetometer coordinate system and the spacecraft (star imager) coordinate system.

where superscript ^{T} indicates matrix transpose. In the main field formalism it is assumed that the field $\underline{B}_G$ is the negative gradient of a potential function, V, where V is expressed as a spherical harmonic series

$$V(\mathbf{r}) = a \sum_{n=1}^{n^*} \sum_{m=0}^{n} \left(\frac{a}{r}\right)^{n+1} [g_n^m \cos(m\phi) + h_n^m \sin(m\phi)] P_n^m(\cos\theta) , \quad (10.38)$$

where fields external to the measurement region are ignored, a is the mean radius of the Earth, the P_n^m are associated Legendre functions of degree n and order m, normalized after the convention of Schmidt, the $\{g_n^m$ and $h_n^m\}$ are model coefficients to be determined, and n^* is a selected truncation level. See Langel (1987) for a discussion of truncation and of main field modeling in general. Equation (10.37) is linear in the $\{g_n^m$ and $h_n^m\}$ but non-linear in the ε_j, necessitating an iterative solution. For simplicity, here measurements of components that are non-linear in the main field parameters, such as field magnitude, declination and inclination, will be ignored. However, they are readily included in the iterative procedure, see, eg., Langel (1987). If the estimate of $(\mathbf{B}_0)_i$ after the kth iteration is $(\mathbf{B}_0)_i^k$ then define

$$(\delta\mathbf{B})_i^k = (\mathbf{B}_0)_i - (\mathbf{B}_0)_i^k . \quad (10.39)$$

Collecting together the $\delta(\mathbf{B})_i^k$ for all i into a $3N$ dimensioned vector $\delta\mathbf{c}$, then as before the model equation for the kth iteration can be written

$$\delta \mathbf{c}^k = D^k \delta \mathbf{p}^k , \qquad (10.40)$$

where the δp_j are the adjustments to be made to the model parameters, ie., to the $\{g_n^m, h_n^m, \varepsilon_j\}$, and the elements of D, D_{ij}, are the partial derivatives of the c_i relative to the p_j. To compute the partial derivatives with respect to the Euler angles,

$$\frac{\partial \mathbf{B}_0}{\partial \varepsilon} = \frac{\partial T_{\mathrm{S/C}}^{\mathrm{T}}}{\partial \varepsilon} T_{\mathrm{GS}}^{\mathrm{T}} \mathbf{B}_{\mathrm{G}} , \qquad (10.41)$$

with

$$\frac{\partial T_{\mathrm{S/C}}^{\mathrm{T}}}{\partial \varepsilon_1} = \begin{pmatrix} (T_{\mathrm{S/C}})_{13} & 0 & -(T_{\mathrm{S/C}})_{11} \\ (T_{\mathrm{S/C}})_{23} & 0 & -(T_{\mathrm{S/C}})_{21} \\ (T_{\mathrm{S/C}})_{33} & 0 & -(T_{\mathrm{S/C}})_{31} \end{pmatrix} , \qquad (10.42)$$

$$\frac{\partial T_{\mathrm{S/C}}^{\mathrm{T}}}{\partial \varepsilon_2} = \begin{pmatrix} \cos\varepsilon_1 (T_{\mathrm{S/C}})_{12} & -\cos\varepsilon_2 & -\sin\varepsilon_1 (T_{\mathrm{S/C}})_{12} \\ \cos\varepsilon_1 (T_{\mathrm{S/C}})_{22} & \cos\varepsilon_3 (T_{\mathrm{S/C}})_{12} & -\sin\varepsilon_1 (T_{\mathrm{S/C}})_{22} \\ \cos\varepsilon_1 (T_{\mathrm{S/C}})_{32} & -\sin\varepsilon_3 (T_{\mathrm{S/C}})_{12} & -\sin\varepsilon_1 (T_{\mathrm{S/C}})_{32} \end{pmatrix} , \qquad (10.43)$$

$$\frac{\partial T_{\mathrm{S/C}}^{\mathrm{T}}}{\partial \varepsilon_3} = \begin{pmatrix} 0 & 0 & 0 \\ (T_{\mathrm{S/C}})_{31} & (T_{\mathrm{S/C}})_{32} & (T_{\mathrm{S/C}})_{33} \\ -(T_{\mathrm{S/C}})_{21} & -(T_{\mathrm{S/C}})_{22} & -(T_{\mathrm{S/C}})_{23} \end{pmatrix} , \qquad (10.44)$$

The partial derivatives with respect to the main field model parameters are

$$\frac{\partial \mathbf{B}_0}{\partial \{g_n^m, h_n^m\}} = T_{\mathrm{S/C}}^{\mathrm{T}} T_{\mathrm{GS}}^{\mathrm{T}} \frac{\partial \mathbf{B}_{\mathrm{G}}}{\partial \{g_n^m, h_n^m\}} . \qquad (10.45)$$

Note on the Choice of Euler Angle Definition

The Euler decomposition of a rotation matrix R is given by:

$$R = R_i(\alpha) R_j(\beta) R_k(\gamma) \qquad (10.46)$$

where the sub-script indicates about which nominal axis the proper rotation is to be carried out. Evidently, the rotation matrix is a function of three free parameters, here being a single rotation angle in each of the three matrix factors, ie. the α, β, and γ. Values of adjacent sub-scripts cannot be equal since that would collapse the factorization,eg.

$$R_j(\beta) R_j(\gamma) = R_j(\beta + \gamma) . \qquad (10.47)$$

This property immediately establishes two classes of Euler decompositions: the $i = k \neq j$ class, denoted E_{iji} and the $i \neq j \neq k$ class, denoted E_{ijk}.

10.1. NON-ABSOLUTE SATELLITE VECTOR MAGNETIC FIELD DATA

Now, consider two complete, orthonormal bases represented by the column vectors of $\mathbf{U}_A$ and $\mathbf{U}_B$ such that $\mathbf{U}_B$ is a slight perturbation of $\mathbf{U}_A$. Specifically, consider a slight perturbation in the ith basis vector of $\mathbf{U}_A$ only, such that:
$$\mathbf{U}_{Bi} \approx \mathbf{U}_{Ai} \,. \tag{10.48}$$
It can be shown that this implies a class E_{ijk} Euler decomposition of:
$$R_j(\beta) \approx R_k(\gamma) \approx I \tag{10.49}$$
$$R \approx R_i(\alpha) \tag{10.50}$$

However, it can also be shown that a class E_{iji} Euler decomposition would render:
$$R_j(\beta) \approx I \tag{10.51}$$
$$R \approx R_i(\alpha) R_i(\gamma) = R_i(\alpha + \gamma) \,. \tag{10.52}$$

Evidently, there is a parameter redundancy in the class E_{iji} decomposition that will lead to aliasing and colinearity in any procedure used to estimate them separately, ie. they are inseparable.

As stated earlier, it is expected that post-launch attitude adjustments will be minor, and so one must deal with the small-scale perturbation regime. Therefore, it is recommended that Euler decompositions of class E_{ijk} be used when determining the three free parameters of $T_{S/C}$. As an example, consider a E_{313} or E_{zxz} decomposition of $T_{S/C}$ given by:

$$T_{S/C} =$$
$$\begin{pmatrix} \cos\varepsilon_3 \cos\varepsilon_1 - \cos\varepsilon_2 \sin\varepsilon_1 \sin\varepsilon_3 & \cos\varepsilon_3 \sin\varepsilon_1 + \cos\varepsilon_2 \cos\varepsilon_1 \sin\varepsilon_3 & \sin\varepsilon_2 \sin\varepsilon_3 \\ -\sin\varepsilon_3 \cos\varepsilon_1 - \cos\varepsilon_2 \sin\varepsilon_1 \cos\varepsilon_3 & -\sin\varepsilon_3 \sin\varepsilon_1 + \cos\varepsilon_2 \cos\varepsilon_1 \cos\varepsilon_3 & \cos\varepsilon_3 \sin\varepsilon_2 \\ \sin\varepsilon_1 \sin\varepsilon_2 & -\sin\varepsilon_2 \cos\varepsilon_1 & \cos\varepsilon_2 \end{pmatrix} \tag{10.53}$$

where ε_1, ε_2, and ε_3 are applied in that order. If the initial and final Z-axes are near coincidental, then $\varepsilon_2 \approx 0$. Furthermore, if $\varepsilon_2^k \approx 0$ for the current estimate of ε_2, then:

$$\frac{\partial \mathbf{B}_0}{\partial \varepsilon_i} = \frac{\partial T_{S/C}^T}{\partial \varepsilon_i} T_{GS}^T \mathbf{B}_G \,, \tag{10.54}$$

with
$$\frac{\partial T_{S/C}^T}{\partial \varepsilon_1} \approx \frac{\partial T_{S/C}^T}{\partial \varepsilon_3} \approx \begin{pmatrix} (T_{S/C})_{21} & (T_{S/C})_{22} & 0 \\ -(T_{S/C})_{11} & -(T_{S/C})_{12} & 0 \\ 0 & 0 & 0 \end{pmatrix} \tag{10.55}$$

and so ε_1 and ε_3 are inseparable. A previous version of the GSFC modeling software used the E_{iji} notation and failed to represent the attitude calibration for Magsat in an adequate manner.

The relation of the euler angles to spacecraft "roll", "pitch", and "yaw" angles is dependent on the spacecraft axis designations. For example, since the Magsat z-axis was designated to be in the along-track direction, ε_2 was in the roll direction, ε_y (radial axis) was yaw, and ε_x (cross track axis) was pitch.

Extended Formalism

This formalism can be extended to include scale factors and transformation from non-orthogonal to orthogonal magnetometer coordinates. Combining (10.31) and (10.35)

$$\mathbf{B}_G = T_{GS} T_{S/C} \mathbf{B}_0 = T_{GS} T_{S/C} H \left[S \mathbf{v}_i + \mathbf{b} + \ldots \right]. \tag{10.56}$$

The model equation then becomes

$$\mathbf{v}_i = T_{SL} T_{NO} T_{SC}^T \mathbf{B}_G + \eta + \ldots, \tag{10.57}$$

where $T_{SL} = S^{-1}$, $T_{NO} = H^{-1}$ and $\eta = -T_{SL}\mathbf{b}$. The extended formalism is incorporated into the GSFC main field modeling software; stability considerations similar to those of the vector-scalar magnetometer comparison apply.

It may seem that, in principle, the extended formalism within main field modeling is more general, and its solution for the scale factors and non-orthogonality more accurate than the solution for those quantities in the simple calibration of the fluxgate magnetometer by comparison with the output of a scalar magnetometer. However, the accuracy of the field model solution depends upon knowledge of spacecraft position and is affected by fields which are not modeled, eg., fields with sources in the ionosphere and lithosphere. When an absolute scalar magnetometer is included on the spacecraft, the best procedure is to first use it to calibrate the vector fluxgate magnetometer and then to use the calibrated data in the field modeling process.

However, there are contexts in which no absolute scalar magnetometer is available on the spacecraft. Several spacecraft have included only a fluxgate magnetometer, with no accompanying scalar instrument, eg., the UARS, POGS and DE-1 spacecraft (Langel et al., 1997) and the DMSP-12 and DMSP-13 spacecraft (Sabaka et al., 1997). In this situation the scalar data used for calibration of the vector instrument, ie., the scalar reference, is computed from the most accurate main field model available, and the accuracy

10.1. NON-ABSOLUTE SATELLITE VECTOR MAGNETIC FIELD DATA

of the result depends upon the accuracy of that model. In this case, the solution for scale factors and non-orthogonality from the extended field model formalism may be more accurate.

10.1.4 Discussion

Two aspects of in-flight calibration of satellite magnetic field data have been discussed in the context of a general data processing scheme. These methods are based on those developed and successfully applied for the Magsat mission. The accuracy of the results depends upon the quality of the magnetometer data, ie., their accuracy and stability, and on the accuracy and stability of the attitude knowledge of the magnetometer. For Magsat, the deviation of the in-flight calibrations from the pre-flight calibration is small and use of the pre-flight calibration as em a priori input to the solution lent stability to that solution. Application of these techniques in a modified form to data of lesser quality (Langel et al., 1997; Sabaka et al., 1997) required substantial regularization to stabilize the solutions. In all applications of the methods, solution stability must be investigated. Only those parameter changes from pre-flight values required by the data should be applied.

Little has been said of data selection. Clearly, the data used in either part of the calibration must be selected so as to occupy as much of the range of the calibration parameters as possible, so that those parameters are resolvable in the fitting procedure. In the case of calibration of a vector instrument using data from a scalar instrument, this implies a sampling along each instrument axis at field levels over the entire instrument range. Torben Risbo (Personal communication, 1995) has found that a data set in which the field magnitude is held constant while the direction samples the entire sphere around the instrument seems to work well in the case of the Ørsted magnetometer. In the case where attitude parameters are included within a solution for Earth's main field, the data must also be sufficiently distributed in geographic space to obtain resolution of the parameters of the main field model. In the scheme of Fig. 10.1, the calibration procedures described in this report comprise Step 2 and Step 4. In this procedure, calibration parameters are adjusted periodically from subsets of the data with results then applied to all of the data. The frequency of determination of the calibration parameters depends upon if and how rapidly they are changing with time. This can only be determined by trial-and-error. Although the procedures generally work well, there may remain some scatter in the solution parameters when they are plotted as a function of time. Fig. 10.5 shows the bias adjustment along one axis of the Magsat fluxgate magnetometer as a function of time and Fig. 10.6 shows the Magsat attitude angle adjustments as a function of time. Scatter is

234   CHAPTER 10. MAGNETOMETER INFLIGHT CALIBRATION

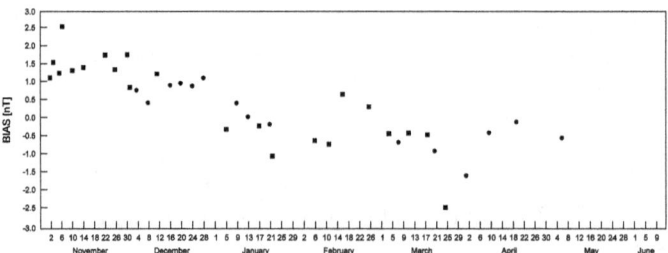

Figure 10.5: Variation in the bias along axis 1 of the Magsat magnetometer as determined by in-flight calibration.

evident in each parameter. For Magsat, each calibration parameter was fit as a function of time, providing a continuous representation of that parameter. Under the assumption that the true variation of the calibration parameters is smooth and continuous, fitting with time smooths out the inevitable scatter in the time variation due to inaccuracies in parameter determination and insures that there are no artificial discontinuities in the data at the calibration epochs.

10.1. NON-ABSOLUTE SATELLITE VECTOR MAGNETIC FIELD DATA

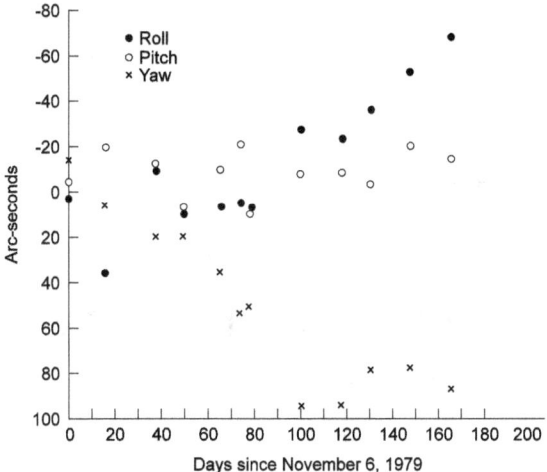

Figure 10.6: Variation in roll, pitch, and yaw axes of the Magsat magnetometer axes, as determined by in-flight calibration.

References chapter 10

1. Lancaster, E.R., T. Jennings, M. Morrissey, and R.A. Langel, Magsat vector magnetometer calibration using Magsat geomagnetic field measurements, NASA GSFC Technical Memorandum 82046, 1980.

2. Langel, R.A., Main Field, Chapter Four in Volume One of Geomagnetism, ed., J.A. Jacobs, Academic Press, 1987

3. Langel, R.A., J. Berbert, T. Jennings, and R. Corner, Magsat data processing: A report for investigators, NASA GSFC Technical Memorandum 82160, 1981.

4. Langel, R.A., J.A. Conrad, T.J. Sabaka, and R.T. Baldwin,Adjustments of UARS, POGS, and DE-1 satellite magnetic field data for modeling of Earth's main field, Jour. Geomagn. Geoelectr., in press, 1997.

5. Sabaka, T.J., J.A. Conrad, J.M.G. Merayo, and R.A. Langel, Analysis of Defense Meteorological Satellite Program 12 and 13 Satellite Magnetometer Measurements, Hughes STX Report, 1997

Chapter 11

Temperature Test Facility for Magnetometers

Werner Magnes and Karl Mocnik

Space Research Institute, IWF, GRAZ

11.1 Introduction

The operation and calibration testing of any sensor at high or low temperatures normally doesn't involve special problems. But if one has to calibrate fluxgate sensors that are used for very small magnetic fields (less than 1 000 nT) the test procedure becomes much more sophisticated. For this case parts of the temperature test facility have to be packed into a magnetic shielding because of the huge Earth Field and the superimposed disturbances.

The temperature test facility was constructed for the MAREMF-OS magnetometer (range: $\pm 128\ nT$) as well as for future spaceborne magnetic field experiments. It has the unique advantage that the sensor under test can be rotated by the Teflon pipe (see Figs. 11.1 and 11.2) for making an accurate offset measurement during the whole temperature cycle. Cooling and heating within a single temperature cycle is not possible because of the mechanical complexity of the entire test equipment. The test device consists of the following main components: three layer magnetic shielding set, low temperature equipment, high temperature equipment and calibration coil.

11.1. INTRODUCTION   237

11.1.1 Magnetic Shielding Set

In order to shield the earthfield sufficiently, a three layer cylinder set consisting of a high permeability iron-sheet is used (Fig. 11.1). It guarantees a shielding factor of about 50 000. The dimensions of the inner cylinder (diameter: 280 mm) allow the insertion of a glass dewar with a volume of about 20 litres. The remaining field in the center of the cylinder is approximately 1 nT, which gives a unique rest field environment for a lot of calibration measurements (e.g.: measuring the offset by turning the sensor, noise and noise density, transfer function, long-term stability etc.).

11.1.2 Low Temperature Equipment

The drawing in Fig. 11.1 shows the configuration of the low temperature test equipment. A PC controlled regulation unit (VECTOTHERM) drives the heater inserted in the liquid nitrogen (LN2) storage tank until the nominal temperature is reached. The actual temperature for the regulation is measured by a Pt-100 element very close to the magnetic field sensor. The vaporized nitrogen is blown through the double walled and evacuated supply pipe into the dewar. The glass dewar isolates the temperature environment around the magnetic field sensor for minimizing temperature losses and temperature changes of the magnetic shielding. Temperatures down to $-150°C$ can easily be obtained, limited only by the test sensor and its thermal energy content. The nominal temperature as well as the cooling velocity (e.g. $0.01°C/sec$) between two temperature values can be adjusted stepwise with a regulation accuracy of approximately $\pm 0.5°C$. The MAREMF-OS qualification model sensor was successfully tested down to $-100°C$ as well as the MAREMF-OS flight model sensor down to $-80°C$.

11.1.3 High Temperature Equipment

Contrarily to the low temperature set-up described above, the heating device requires a hot air circulation. This is realized by a fan-heater system which

Chapter 11. Temperature Test Facility for Magnetometers

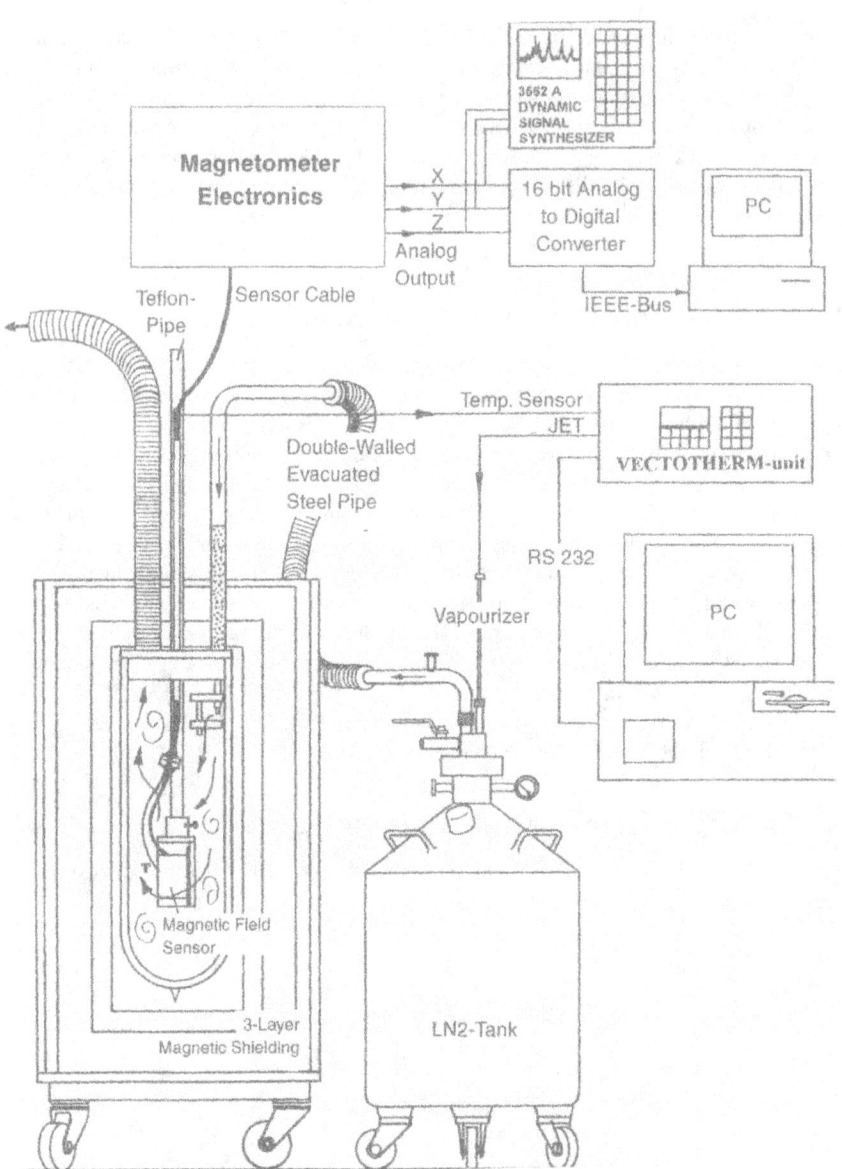

Figure 11.1: Low temperature test configuration.

11.1. INTRODUCTION

is connected to the glass dewar via two flexible and heat resistant tubes (see Fig. 11.2). The VECTOTHERM regulation unit controls the power of the heater in Fig. 11.2 in the same way as the heater within the liquid nitrogen tank described before. Temperatures up to

$$150°C$$

can be reached. The MAREMF-OS qualification model was successfully tested up to

$$100°C$$

11.1.4 Calibration Coil

A cylindrical calibration coil is placed between the glass dewar and the inner magnetic shielding cylinder for applying test fields to the sensor during the measurement cycle. At the top and the bottom of the coil the grid pitch is manufactured smaller than in the middle in order to smooth out the field strength variation. The coil coefficient is 1.18 $nT/$.

11.1.5 Test Result

As one example for the high and low temperature tests the offset drift results of the MAREMF-OS flight model are presented in Fig. 11.3. The offset drift over the temperature range from $-80°C$ up to $70°C$ is less than 2 nT. Approximately the same offset drift was measured with the spare and qualification model.

240 CHAPTER 11. TEMPERATURE TEST FACILITY FOR MAGNETOMETERS

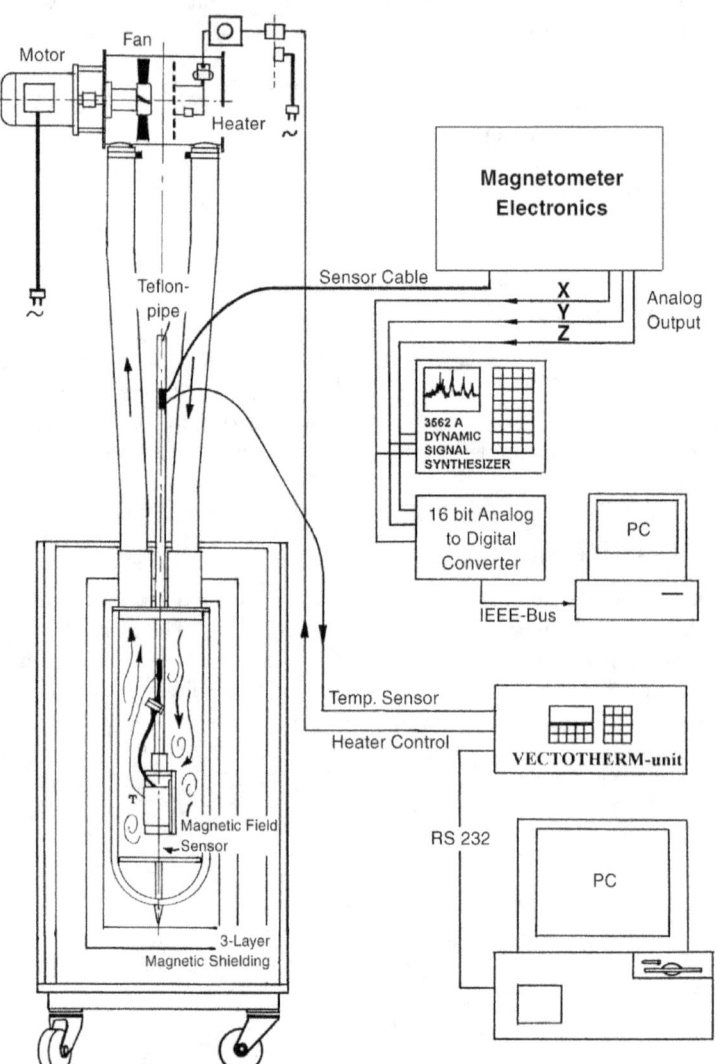

Figure 11.2: High temperature test configuration.

11.1. INTRODUCTION

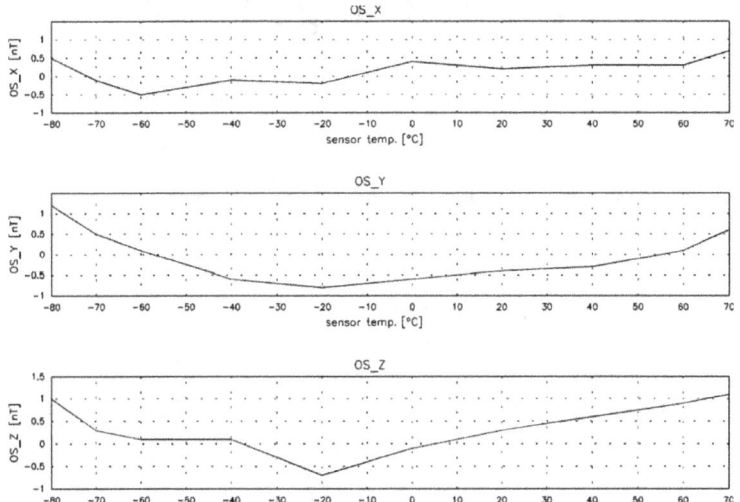

Figure 11.3: Offset drift versus sensor temperature (MAREMF-OS FM)

Chapter 12

General Fluxgate Magnetometer Bibliography

compiled by:

F.Primdahl[1], Yu.V.Afanassiev[2], F.Kuhnke[3] and R.C.Snare[4].

[1] Danish Space Research Institute Lyngby, Copenhagen
[2] MAG – Sensors, St.Petersburg,
[3] formerly Institute of Geophysics, Technical University Braunschweig
[4] UCLA Institute of Geophysics and Planetary Physics, Los Angeles

1. Primdahl, F., Bibliography of Fluxgate Magnetometers, Publications of the Earth's Physics Branch, Vol. 41 - No. 1, pp 1-14, Department of Energy, Mines and Resources, Ottawa, Canada, 1970.

2. Serson, P.H. and F. Primdahl, Bibliography of Magnetometers, Publications of the Earth's Physics Branch, Vol. 43 - No. 8, pp 501-506, Department of Energy, Mines and Resources, Ottawa, Canada, 1972.

3. Coles, R.L. (ed.), A Bibliography of Magnetometers, Proceedings of the International Workshop on Magnetic Observatory Instruments, Geological Survey of Canada, 1988.

4. Afanasenko, M. P. and R. Ja. Berkman, O Fizicheskoi Osushchestvimocti Zadannogo Vida Kharakteristiki Preobrazovanija Ferrozonda v Neodiorodnom Izmerjaemom Magnitnom Pole (The Physical Feasability of Securing a

Specified Form of Ferroprobe Conversion Characteristic for making measurements in a Nonuniform Magnetic Field), Defektoskopia (J. Non-Destructive Testing), No. 2, 63 - 70, 1969.

5. Afanasenko, M. P. and R. Ja. Berkman, K Teorij Ferrozondov Rabotajushchikh v Neodnorodnikh Magnitnikh poljakh (title in English), Otbor i Peredacha Informatsij, Vol. 22, 69 - 73, 1969.

6. Afanasenko, M. P. and R. Ja. Berkman, Sravnitel'nij Analiz Shumovikh Protsessov v Magnitnikh moduljatorakh i kol'tsevikh Ferrosondakh (Comparison of Noise Processes in Magnetic Modulators and Ring Ferroprobes), Avtometria (Automatic Measurement), No. 3, 72 - 81, 1972.

7. Afanasenko, M. P., P. Ja. Berkman, B. L. Bondaruk and A. V. Itskovich, Mnogokomponentnij Izmeritel' Napjazhennocti Magnitnigo Polja Kol'tsevim Ferrozondom (Multiple Component Magnetic field Intensity Meter with a Ring Ferromagnetic Probe), Otbor i Peredacha Informatsij, Vol. 41, 92 - 98, 1974.

8. Afanassiev, Yu. V., Perspectives of Lovering the Noise Level of Fluxgate Sensors, (I have an English translation by Pavel Ripka; but no data on the original paper in russian).

9. Arnold, H., A. Weixler, G. Berghofer, K. Schwingenschuh, B. Snare and D. Pierce, Development of the Magnetometer Analog Electronics for the MAREMF Experiment, Report IWF-9005, Institut für Weltraumforschung, Steyrergasse 17 - 19, A - 8010 Graz, Austria, 1990.

10. Auster, H. U., A. Lichopoj, J. Rustenbach, H. Bitterlich, K. H. Fornaçon, O. Hillenmaier, R. Krause, H. J. Schenk and V. Auster, Concept and First Results of a Digital Fluxgate Magnetometer, Meas. Sci. Technol., Vol. 6, 477 - 481, 1995.

11. Balogh, A., T. J. Beek, R. J. Forsyth, P. C. Hedgecock, R. J. Marquedant, E. J. Smith, D. J. Southwood and B. T. Tsurutani, The Magnetic Field Investigation on the Ulysses Mission: Instrumentation and Preliminary Scientific Results, Astron. & Astrophys. Suppl. Series, Vol. 92, 221 - 236, 1992.

12. Balogh, A., S. W. H. Cowley, M. W. Dunlop, D. J. Southwood, J. G.Thomlinson, K. H. Glassmeier, G. Musmann, H. Luehr, M. H. Acuna, D. H. Fairfield, J. A. Slavin, W. Riedler, K. Schwingenshuh, F. M. Neubauer, M. G. Kivilson, R. C. Elphic, F. Primdahl, A. Roux, and B. T. Tsurutani, The Cluster Magnetic Field Investigation: Scientific Objectives and Instrumentation, ESA SP-1159, 95 - 114, March 1993.

13. Berkman, R. Ja., On Increasing the Accuracy of Geomagnetic Measurements by means of Saturated Core Probes, Dopovidi Akademi Nauk Ukrainckoi RSR, Geofizika, No. 4, 350 - 353, 1957.

14. Berkman, R. Ja., About a New Type of Magnetomodulation Transducers for Measurement of the Magnetic Field Gradient (in Russian), Avtom. Kontrol i Izmer. Tech., Vol. 4, 157 - 162, 1960.

15. Berkman, R. Ja., R. E. Martinjuk-Lototskij, V. N. Mikhailovskij and Ju. I. Spektor, O Stabil'nosti Nula Magnitomoduljationnikh Datchikob Rabotajushchikh v Neodinorodnikh magnitikh Poljakh (title in English?), Avtomaticheskij Kontrol' i Izmeritel'naja Tekhnika, Vol. 8, 90 - 94, 1964.

16. Berkman, R. Ja., R. E. Martinjuk-Lototskij and Ju. I. Cpektor, Osobennosti Rascheta Ferrozondov s Kol'tsevimi Serdechnikami (title in English?), Avtomaticheskij Kontrol' i Izmeritel'naja Tekhnika, Vol. 8, 95 - 99, 1964.

17. Berkman, R. Ja. and V. V. Belousko, K Voprosu o Kharakteristike Preobpazovanija Ferrozonda v Rezhime Izmerenija Neodnorodnikh Magnitnikh Polei (Conversion Characteristics of a Ferro-Probe when Mesuring Nonuniform Magnetic Fields), Defektoskopia (J. Non-Destructive Testing), No. 5, 61 - 67, 1965.

18. Berkman, R. Ja., The Effect of Higher Even Harmonics in the Excitation Circuit of Magnetic Modulators, Translated from: 'Avtomatika i Telemekhanika', Vol. 26, No. 2, pp 384 - 387, 1965.

19. Berkman, R. Ja., O Stabil'nosti Chuvstvitel'nosti Magnitnikh Moduljatorov Vtoroi Garmoniki v Zavisimosti ot Rezhima Raboti Vizbuzhdajushchei i Izmeritel'noi Tsepi, Voproc'i Teorij Electricheskikh Tsepei Dlja Preobrazovanija Izneritel'noi Informatsij, Kiev: Naukova Dumka, pp. 69 - 76, 1967.

20. Berkman, R. Ja. and B. L. Bondaruk, Magnitnij moduljator s Visokoi Chuvstbitel'nostju po Toku (Title in English?), Otbor i Peredacha Informatsij, Vol. 26, 78 - 83, 1970.

21. Berkman, R. Ja., B. L. Bondaruk and V. M. Fedotov, Optimalnaja Skhema Vozbuzhdenija Nizkopogovikh Magnito-Moduljatsionnikh Preobrazovatelei i Analiz ee rabotij (Title in English?), Otbor i Peredacha Informatsij, Vol. 32, 43 - 51, 1972.

22. Berkman, R. Ja., B. L. Bondaruk and V. M. Fedotov, Analiz Chuvstvitel'nosti Magnitnikh Moduljatorov v Ferrorezonansnom Rezhime Vozbuzhdenija , Otbor i Peredacha Informatsij, Vol. 43, 60 - 64, 1975.

23. Berkman, R. Ja. and L. I. Rakhlin, Analiz Pogreshnostei ot Chetnikh Garmonik v Ferrorezonansnom rezhime vozbyzhdenija magnitnikh moduljatorov (title in English ?), Otbor i Peredacha Informatsij, Vol. 49, 62 - 68, 1976.

24. Boll, R., and G. Hinz, Sensors of Amorphous Metal, Technisches Messen tm (Germany), Vol. 52, 189 - 198, (in German), 1985.

25. Bornhöfft, W. and G. Trenkler, Magnetic Field Sensors: Flux Gate Sensors, in: Sensors, A Comprehensive Survey, eds.: W. G?pel, J. Hesse and J.N. Zemel, Vol. 5, Magnetic Sensors, Ch. 5, pp 153 - 203, VCH, 1989.

26. Bosum, W., D. Eberle and H.-J. Rehli, A Gyro-Oriented 3-Component Borehole Magnetometer for Mineral Prospecting with Examples of its Application, Geophysical Prospecting, Vol. 36, 933 - 961, 1988.

27. Brekke, A., T. Hansen, S. Berger, T. Brattli and B. Holmslet, Upgraded Magnetic Observatories in the Arctic, Phys. Earth & Planet. Interiors, Vol. 59, 89 - 96, 1990.

28. Carter, M., Experimental Investigation of a Recent Fluxgate Theory, M. Sc. Thesis, The University of British Columbia, Department of Geophysics and Astronomy, November 1988.

29. Dalpadado, R. N. G., Bipolar Magnetometer Sensor Similar to Fluxgate Device But Having Equal Input and Output Sinusoidal Frequences and Zero Barkhausen Noise, Electronics Letters, Vol. 28, 1662 - 1663, 1992.

30. Engelter, A., A Fluxgate Magnetometer with a Metallic Glass Core, IEEE Trans. Mag., MAG-22, 299 - 300, 1986.

31. Farthing, W. H., M. Sugiura, B. G. Ledley and L. J. Cahill, Jr., Magnetic Field Observations on DE-A and B, Space Sci. Instrum., Vol. 5, 551 - 560, 1981.

32. Fornaçon, K. H., M. Müller and I. Fischer, A Fluxgate System for Magnetic Field Measurement in Space, J. Magnetism Magn. Materials, 1994.

33. Fowler, J. Thomas, "New Technology" Magnetic Attitude Sensors in Autonomous Underwater Vehicle (AUV) Applications, 1990 IEEE Symposium on Autonomous Underwater Vehicle Technology, Washington, June 5-6, 1990.

34. Gao, Z.-C. and R. Doncaster Russell, Fluxgate Sensor Theory: Sensitivity and Phase Plane Analysis, IEEE Trans. Geosci. Remote Sens., GE-25, 862 - 870, 1987.

35. Garner, H. D., Improved Flux-gate Magnetometer, Applied Optics, Vol. 27, p. 1366, 1988.

36. Geyger, W. A., Second-harmonic-type Flux-gate Magnetometers, in 'Magnetic-amplifier Circuits', 17.4, pp 271 - 278, McGraw-Hill Inc., 1957.

37. Ghatak, S. K., and A. Mitra, A Simple Fluxgate magnetometer Using Amorphous Metal Alloys, Journ. Magnetism Mag. Materials, Vol. 103, 81 - 85, 1992.

38. Harada, K., Y. Kozima, Y. Fujii and S. Takeuchi, Measurement of Blast Furnace Inner-State by Magnetometer, IEEE Trans. Mag., MAG-16, 698 - 700, 1980.

39. Heyningen, M., The Evolution of the Modern Electronic Compass, NMEA News, pp 13 - 77, January,February, 1987.

40. Huan, Yinn-Nien, On the Digital Geomagnetic Observatory at Lunping, Taiwan, Phys. Earth & Planet. Interiors, Vol. 59, 66 - 77, 1990.

41. Ito, H. and Matsumoto, K., The Development of a Field Magnetometer (in Japanese), Mitsubishi Denki Giho (Japan), Vol. 61, 81 - 86, 1987.

42. Kawahito, S., Y. Sasaki, H. Sato and T. Nakamura, Micromachined Solenoids for Highly Sensitive Magnetic Sensors, 1991 Int. Conf. Solid State Sensors and Actuators, IEEE, pp. 1077 - 1080, 1991.

43. Kawahito, S., Y. Sasaki, H. Sato, T. Nakamura and Y. Tadokoro, A Fluxgate Magnetic Sensor With Micro-Solenoids and Electroplated Permalloy Cores, Sensors and Actuators A, Vol. 43, 128 - 134, 1994.

44. Kim, H. C. and Chung Sam Jun, A New Method for Fluxgate Magnetometers Using the Coupling Property of Odd and Even Harmonics, Meas. Sci. & Technol., Vol. 6, 898 - 903, 1995.

45. Kivelson, M. G., K. K. Khurana, J. D. Means, C. T. Russell and R. C. Snare, The Galileo Magnetic Field Investigation, Space Sci. Rev., Vol. 60, 357 - 383, 1992.

46. Koizumi, K., J. Segawa, H. Toh, J. L. Oubina Carretero and Y. Tanaka, Simultaneous Measurements and their Comparison at the Sea Floor Using a Fluxgate Vector Magnetometer and a Proton Scalar Magnetometer, J. Geomagn. Geoelectr., Vol. 41, 491 - 506, 1989.

47. Kolachevskij, N. N., in: "Fluktuatsionnie Javlenija v Ferromagnitnikh Materialakh, Glava 10, Magnitnie Shumi v Chetnogarmonichskikh Ferromagnitnikh Preobrazovateljakh i Drugikh Priborakh s Ferromagnitnimi Serdechnikami", (English translation of the title ?) pp 126 - 150, Moskva "Nauka" Glavnaja Redaktsija Fisiko-Matematicheskoi Literaturi, 1985.

48. Krishna Rao, D. A. V., Underwater Magnetic Survey, Defence Science Journ. (India), Vol. 37, 319 - 325, 1987.

49. Kuwashima, M., Accuracy in Geomagnetic Measurements of Japanese Magnetic Observatories, Phys. Earth & Planet. Interiors, Vol. 59, 104 - 111, 1990.

50. Lancaster, D., A Solid-State Digital Compass, Radio-Electronics, Vol. 59, 33 - 36, 1988.

51. Lauridsen, E. Kring, Experiences with the DI-Fluxgate Magnetometer Inclusive Theory of the Instrument and Comparison with Other Metods, Danish Meteorological Institute, Geophysical Papers, R-71, 1981.

52. Lepping, R. P., M. H. Acuña, L. F. Burlaga, W. M. Farrell, J. A. Slavin, K. M. Schatten, F. Mariani, N. F. Ness, F. M. Neubauer, Y. C. Whang, J. B. Byrnes, R. S. Kennon; P. V. Panetta, J. Scheifele and E. M. Worley, The Wind Magnetic Iield Investigation, Space Science Review, Vol. 71, 207 - 229, 1995.

53. Lenz, J. E., A Review of Magnetometer Sensors, Proc. IEEE, Vol. 78, 1990.

54. Liu Shi-jie, Lu Jun, Ma Lian-yan, Tang Jing-bo and Huo Guan-qun, Development and Application of Three Component High Resolution Fluxgate Magnetometer CTM-302 (in Chinese), Acta Geophysica Sineca, Vol. 33, 566 - 576, 1990.

55. Liu, S. W., C. X. Liu and Z. Ding, Finite Element Computer-Aided Optimal Design of the Magnetic Field of Fluxgate Magnetometers, Proc. IECON '89, 15'th Annual Conf. IEEE Industrial Electronics Soc., Vol. 4, pp 817 - 822, IEEE, 1989.

56. Liu, S. W., Z.-N. Zhiang and J. C. Hung, A High Accuracy Magnetic Heading System Composed of Fluxgate Magnetometers and a Microcomputer, Proc. IECON '89, 15'th Annual Conf. IEEE Industrial Electronics Soc., Vol. 4, pp 148 - 152, IEEE, 1989.

57. Lühr, H., N. Klocker, W. Oelschlagel, B. Hasler and M. Acuña, The IRM Fluxgate Magnetometer, IEEE Trans. Geosci. Rem. Sens., Vol. GE-23, 259 - 261, 1985.

58. Mariani, E., M. Candidi and R. Terenzi, Temag Experiment for the Tethered Satellite System, NASA, Report no.: IFSI-85-14, ESA-86-97098, 41p, 1985.

59. McIntyre, Steven A., Magnetic Field Sensor Design, Sensor Review, Vol. 11, 7 - 12, 1991.

60. McPherron, R. L. and R. C. Snare, A Procedure for Accurate Calibration of the Three Sensors in a Vector Magnetometer, IEEE Trans. Geosci., GE-16, 174 - 177, 1978.

61. Mermelstein, M. D., Magnetoelastic Amorphous Metal Fluxgate Magnetometer, Electronics Letters, Vol. 22, 525 - 526. 1986.

62. Mermelstein, M. D., C. Askins and A. Dandridge, Stress-Relieved Magnetoelastic Amorphous Metal DC Magnetometer, Electronics Letters, Vol. 23, 280 - 281, 1987.

63. Moldovanu, B.-O., C. Moldovanu and A. Moldovanu, Computer Simulation of the Transient Behaviour of a Fluxgate Magnetometric Circuit, EMMA'95, Vienna, 1995.

64. Moldovanu, C. and B.-O. Moldovanu, Simulation of the Fluxgate Sensor Operation Using the Jiles-Atherton Model for the Magnetization Curve Implemented on the PSPICE Computer Program, SMP'95, Vienna, 1995.

65. Möhlmann, D., J. Rustenbach, H.-U. Auster, H. Bitterlich, K.-H. Fornacon, G. Hunsalz, J. Kersten, A. Lichopoi, M. Michaelis, A. Pauli, Th. Roatsch H.-J. Schenk, R. Schr?ter, H. Studemund, The Fluxgate Magnetometer Aboard the PHOBOS Spacecraft, IKF-Preprint 3/90, Institut für Kosmosforschung, Berlin, 1990.

66. Narod, B. B. and J. R. Bennest, Ring-Core Fluxgate Magnetometers for use as Observatory Variometers, Phys. Earth and Planet. Interiors, Vol. 59, 23 - 28, 1990

67. Narod,B. B. and R. D. Russel,Steady-state Characteristics of the Capacitively loaded Fluxgate Sensor,IEEE Transaction on Magnetics,Mag-20,592-597,1984

68. Nielsen Otto V., J. Gutierrez, B. Hernando and Howard T. Savage, A New Amorphous Ribbon Fluxgate Sensor Based on Torsional-Creep-Induced Anisotropy, IEEE Trans. Mag., MAG-26, 276 - 280, 1990.

69. Nielsen, O.V., B. Hernando, J.R. Petersen and F. Primdahl, Miniaturisation of Low-Cost Metallic Glass Flux-Gate Sensors, J. Magnetism and Magnetic Materials, Vol. 83, 405 - 406, 1990.

70. Nielsen, O. V., J. R. Petersen, B. Hernando, J. Gutierrez and F. Primdahl, Metallic Glasses for Fluxgate Applications, Anales de Fisica, B86, 271 - 76, 1990.

71. Nielsen, O. V., J. R. Petersen, A. Fernando, B. Hernando, P. Spisak, F. Primdahl and N. Moser, Analysis of a Fluxgate Magnetometer Based on Metallic Glass Sensors, Measurement Science & Technology, 2, 435 - 440, 1991.

72. Nielsen, O. V., T. Johansson, J. M. Knudsen and F. Primdahl, Possible Magnetic Experiments on the Surface of Mars, J. Geophys. Res. (Planets), Vol. 97, 1037 - 1044, 1992.

73. Nielsen, O. V., F. Primdahl, J. R. Petersen and B. Hernando, Selection and Processing of Metallic Glass Materials for Fluxgate Applications, III International Workshop on Non-Chrystalline Solids, Universidad de Sevilla (Spain), Nov. 5.-8., 1991, in: Trends in Non-Crystalline Solids, Eds.: A. Conde and M. Mill?n, World Scientific Publ. Co., Singapore, p. 417 - 420, 1992.

74. Nielsen, O. V., B. Hernando, G. Herzer and P. Spisak, Nanocrystalline Materials as the magnetic Core in High performance Fluxgate Sensors, NATO Workshop on Nanomagnetic Devices, Madrid Sept. 1992, in: A. Hernando (ed.), Nanomagnetism, NATO ASI Series E: Applied Sciences, Vol. 247, 27 - 31, Kluwer Academic, 1993.

75. Nielsen, O. V., J. R. Petersen, F. Primdahl, P. Brauer, B. Hernando, A. Fernandez, J. M. G. Merayo, and P. Ripka, Development, Construction and Analysis of the "ØRSTED" Fluxgate Magnetometer, Meas. Sci. Technol., Vol. 6, 1099 - 1115, 1995.

76. Noble, R., Fluxgate Magnetometry, Design, Electronics World + Wireless World, 726 - 732, September, 1991.

77. Ondoh, T., S. Kokubun, M. Abe, M. Kuwashima and H. Yamada, Development of SEM Magnetometer for Geostationary Satellite, Rev. Comm. Res. Lab. (Japan), Vol. 35, 451 - 464, 1989.

78. Petersen, J. R., F. Primdahl, B. Hernando, A. Fernandez and O. V. Nielsen, The Ring-Core Fluxgate Sensor Null Feed-Through Signal, Meas. Sci. & Technol., 3, 1149 - 1154, 1992.

79. Piil-Henriksen, J., J. M. G. Merayo, O. V. Nielsen, H. Petersen, J. Raagaard Petersen and F. Primdahl, Digital detection and Feddback Fluxgate Magnetometer, Meas. Sci. Technol., (submitted, Dec., 1995).

80. Player, M. A., Parametric Amplification in Fluxgate Sensors, J. Phys. D: Appl. Phys., Vol. 21, 1473 - 1480, 1988.

81. Pollock, N., Electronic Compass Using a Fluxgate Sensor, Wireless World, pp 49 - 54, October, 1982.

82. Potemra, T. A., L. J. Zanetti and M. H. Acuna, The AMPTE CCE Magnetic Field Experiment, IEEE Trans. Geosci. Rem. Sens., Vol. GE-23, 246 - 249, 1985.

83. Primdahl, F., J. R. Petersen, C. Olin and K. Harbo Andersen, The Short-Circuited Fluxgate Output Current, J. Phys. E.: Sci. Instrum., Vol. 22, 349 - 354, 1989.

84. Primdahl, F., B. Hernando, O. V. Nielsen and J. R. Petersen, Demagnetising Factor and Noise in the Fluxgate Ring-Core Sensor, J. Phys. E.: Sci. Instrum., Vol. 22, 1004 - 1008, 1989a.

85. Primdahl, F., P. A. Jensen and Chr. Boe, Magnetic Measurements With Fluxgate Magnetometer in Rocket Flight, Report DRI-4-90, 26p, Danish Space Research Institute, Denmark, 1990.

86. Primdahl, F., P. Ripka, J. R. Petersen and O. V. Nielsen, The Sensitivity Parameters of the Short-Circuited Fluxgate, Meas. Sci. and Technol., 2, 1039 - 1045, 1991.

87. Primdahl, F., H. Lühr and E. Kring Lauridsen, The Effect of Large Uncompensated Transverse Fields in the Fluxgate Magnetic Sensor Output, DRI 1-92, Danish Space Research Institute, January 1992.

88. Primdahl, F., The Short-Circuited Fluxgate, Slaboproud? Obzor (Electronic Horizon), (Czechoslovakia), 53, 95 - 97, 1992.

89. Primdahl, F., B. Hernando, J. R. Petersen, and O. V. Nielsen, Digital Detection of the Fluxgate Sensor Output Signal, Meas. Sci. Technol., Vol. 5, 359 - 362, 1993.

90. Primdahl, F., O. V. Nielsen, J. R. Petersen, and P. Ripka, High Frequency Fluxgate Sensor Noise, Electronics Lett., Vol. 30, 481 - 482, 1994.

91. Primdahl, F.,The Fluxgate Magnetometer,Journal of Physics E:Scientific Instruments ,Vol.20.637-642,1979.

92. Primdahl, F. and P. Anker Jensen,Noise in the Tuned Fluxgate,Journal of Physics E: Scientific Instruments,Vol.20, 637-642,1987.

93. Rasmussen, O. and E. Kring Lauridsen, Improving Baseline Drift in Fluxgate Magnetometers Caused by Foundation Movements, Using Band Suspended Fluxgate Sensors, Phys. Earths Planet. Inter., Vol. 59, 78 - 87, 1990.

94. Rasmussen, O., Tilt-Compensation of Fluxgate Magnetometers, Geophysical Transactions, Vol. 36, 261 - 269, 1991.

95. Rao, B.V.G.K.V.B.V.P., D. A. V. Krishna Rao and G. V. R. S. Kumar, Peak Height Fluxgate magnetometer for measurement of Small magnetic fields, Defence Science Journ. (India), Vol. 36, 265 - 271, 1986.

96. Rich, F. J., Fluxgate Magnetometer (SSM) for the Defence Meteorological Satellite Program (DMSP) Block 5D-2, Flight 7, Air Force Geophysics Lab, Hanscom AFB, MA, Report No.: AFGL-TR-84-0225, AFGL-IP-326, 30p, 1984.

97. Ripka, P., Contribution to the Theory of Toroidal Ferromagnetic Probe (in Czech), Sdelovaci Technika (Czech Republ.), Vol. 35, 101 - 104, 1987.

98. Ripka, P., F. Jires and M. Machacek, Fluxgate Sensor with Increased Homogeneity, IEEE Trans. Mag., MAG-26, 2038 - 2841, 1990.

99. Ripka, P., Improved Fluxgate for Compasses and Position Sensors, J. Magn. Magn. Materials, Vol. 83, 543 - 544, 1990.

100. Ripka, P., Review of Fluxgate Sensors, Sensors and Actuators A, Vol. 33, 129 - 141, 1992.

101. Ripka, P., Measurement of the Fluxgate Noise Spectrum, IHECO TC-7 Conference: "Measuring Instruments for Low Frequencies", Vienna, March, 1992.

102. Ripka, P., Race-track Fluxgate Sensors, Sensors and Actuators A, Vol. 37 - 38, 417 - 421, 1993.

103. Ripka, P., K. Draxler and P. Kaspar, Race-Track Fluxgate Gradiometer, Electronics Letters, Vol. 29, 1193 - 1194, 1993.

104. Ripka, P., J. R. Petersen, H. Petersen, O. V. Nielsen, and F. Primdahl, Satellite Magnetometer, J. of Electrical Engineering (ISSN 0013-578X), (Elektrotechnicky Casopis), Vol. 45, 107 - 110, 1994.

105. Ripka, P., Magnetic Sensors for Industrial and Field Applications, Sensors and Actuators A, Vol. 41 - 42, 394 - 397, 1994.

106. Ripka. P., J. R. Petersen, H. Petersen, O. V. Nielsen and F. Primdahl, Satellite Magnetometer, J. Electrical Engineering (Electrotechnicky Casopis, Slovakia), Vol. 45, 107 - 110, 1994.

107. Ripka, P., F. Primdahl, O. V. Nielsen, J. R. Petersen and A. Ranta, A. C. Magnetic-Field Measurement Using the Fluxgate, Sensors and Actuators A, Vol. 46-47, 307 - 311, 1995.

108. Russel,R. D.,B.Narod and F. Kollar,Characteristics of the Capacitively Loaded Fluxgate Sensor,IEEE Transactions on Magnetics, MAG-19,126-130,1983.

109. Sa, A. de and K. Heron, A Three-axes Fluxgate Magnetometer With Mulitplexed Sensors, Meas. Sci. Technol., Vol. 4, 633 - 634, 1993.

110. Scott, J. H., G. G. Olson, Three-Component Borehole Magnetometer Probe for Mineral Investigations and Geological Research, Trans. SPWLA Annual Logging Symposium, Soc. Prof. Well Log Analysts, 16p, 1985.

111. Seitz, T., Fluxgate Sensor in Planar Microtechnology, Sensors and Actuators A (Physical), Vol. A22, 799 - 802, 1990.

112. Serson,P. H.and W. L. W. Hannaford,A Portable Electrical Recordingf Magnetometer, Canadian journal of Technology,Vol.34,232-243,1956.

113. Singer, H. J., W. P. Sullivan, P. Anderson, F. Mozer, P. Harvey, J. Wygant and W. McNeil, Fluxgate magnetometer on the CRRES, J. Spacecraft and Rockets, Vol. 29, 599 - 601, 1992.

114. Son, D., A New Type of Fluxgate Magnetometer Using Apparent Coercive Field Strength Measurement, IEEE Trans. Mag., MAG-25, 3420 - 3422, 1989.

115. Son, D., A New Type of Fluxgate Magnetometer for Low Magnetic Fields, Physica Scripta, Vol. 39, 535 - 537, 1989.

116. Sonoda, T., R. Ueda, H. Ikemoto, K. Kudo and K. Kajiwara, Differentially DC Biased Type Magnetic Field Sensor of High Sensitivity, IEEE Trans. Mag., MAG-25, 3396 - 3398, 1989.

117. Southwood, D. J., W. A. C. Mier-Jedrzejowicz and C. T. Russell, The Fluxgate Magnetometer for the AMPTE UK Subsatellite, IEEE Trans. Geosci. Rem. Sens., Vol. GE-23, 301 - 304, 1985.

118. Stewart, B. G., J. A. Rollo and J. F. L. Simmons, Magnitude Determination of Magnetic Fields From Three Orthogonal Measurements, J. Phys. E: Sci. Instrum., Vol. 22, 457 - 461, 1989.

119. Trigg, D. F. and D. G. Olson, Pendulously Suspended Magnetometer Sensors, Rev. Sci. Instrum., Vol. 61, 2632 - 2636, 1990.

120. Tumanski, S., Metrological Properties of Flux-Gate Sensors and Magnetometers (in Polish), Rozprawy Electrotechniczne (Poland), Vol. 32, 813 - 834, 1986.

121. Wang, Zenghua, Characteristic Analysis of the Transducer of a Fluxgate Magnetometer, J. Shanghai Jiatong Univ., Vol. 26, 91 - 98, 1992.

122. Zanetti, L. J., T. Potemra, R. Erlandson, P. Bythrow, B. Anderson, A. Lui, S.-I. Ohtani, G. Fountain, R. Henshaw, B. Ballard, D. Lohr, J. Hayes, D. Holland, M. Acuna, D. Fairfield, J. Slavin, W. Baumjohann, M. Engebretson, K.-H. Glassmeier, G. Gustafsson, T. Iijima, H. Luehr, and F. Primdahl, Magnetic Field Experiment on the Freja Satellite, Space Science Reviews, Vol. 70, pp. 465 - 482, 1994.

123. Zaveta, K., O. V. Nielsen and K. Jurek, A Domain Study of Magnetisation Processes in a Stress-Annealed Metallic-Glass Ribbon for Fluxgate Sensors, J. Magnetism Magn. Materials, Vol. 117, 61 - 68, 1992.

Chapter 13

APPENDIX
Guidelines for Magnetic Cleanliness

compiled by

G.Musmann

with contributions from:

A.Balogh, C.Carr, B.Gramkow, H.Kügler, W.Magnes, K.Mehlem, M.Menvielle, G.Musmann, A.Stadelmann(Neuhaus), F.Primdahl, K.Schwingenschuh, J.Stadelmann, C.Stuntebeck

Abstract

This document is based on the experience in ESA and in European Institutes which has been built up during three decades, including many projects like GEOS, ISEE-B, GIOTTO, TETHER, ULYSSES, CLUSTER, CASSINI-HUYGENS and ROSETTA.

Powerful and efficient test facilities have been developed and used like the MFSA at IABG and the Mobile Coil Facilities called MCF.

In parallel a unique modeling method and a related software has been developed at ESTEC (K.Mehlem: Multiple Dipole Modeling, MDM, later called MAGNET) and successfully used for all projects.

This document deals with the importance of keeping the background magnetic field in the spacecrafts low in order to allow the DC-magnetic experiments to reliably observe the magnetic field.

Based on the cumulative experience from previous space projects available in the international community of space magnetometry a range of different sources of spacecraft magnetic fields is identified and recommendations to reduce their contributions to the magnetic perturbations at the magnetometer sensor(mostly two) are explained and exemplified.

The interaction between the flight hardware suppliers and the EMC review board is seen as a dialogue converging towards the best mutually agreeable solution getting the very best science from the TBD spacecraft program.

Summary

Most important for obtaining magnetically acceptable spacecrafts are the collaborative efforts of all groups involved to keep the source of contaminating magnetic fields reasonably low. The first prerequisite for this is for all groups delivering hardware to test all candidate parts for magnetism as early in the design process as possible and go to the EMC board for help and advice, when potentially problematic magnetic parts must go into the systems.

To help meeting the required magnetic specifications , this document gives a large number of specifications, recommendations and advice on magnetic cleanliness compiled from many sources and drawing on the experience of people involved in similar tasks like.

Chapter 13. Appendix Guidelines for Magnetic Cleanliness

Contents

13.1.0 Introduction
13.2.0 Design Approach
13.2.1 General
13.2.2 Approach to Minimizing Fields from Magnetic Materials
13.2.3 Approach to Minimizing Stray Fields from Electrical Currents
13.2.4 Approach to Minimizing Fields from Eddy Currents
13.2.5 Magnetic Moment Management
13.2.6 Shielding Problems
13.2.7 Approval Process
13.3.0 Achievement of Magnetic Cleanliness
13.3.1 Permanent Fields and Fields from Soft Magnetic materials
13.3.1.1 Materials
13.3.1.2 Electronic Parts
13.3.2 Stray Fields produced by Electric Currents
13.3.2.1 Guidelines
13.3.2.2 Stray Fields of Parts with Magnetic cores
13.3.2.3 Solar Array
13.3.2.4 Special Problem Assemblies
13.3.2.5 Fields from Eddy Currents in Structural Components
13.4.0 Magnetic testing of Flight Hardware
13.4.1 Testing at Assembly and Unit Level

Appendix A
Relations to other units
Appendix B
General Guidelines for the Design of Magnetic Circuit for Magnetically Clean Spacecraft
References chapter 13

13.1 Introduction

The magnetometer investigations on Spacecrafts either near Earth ,Planetary and Interplanetary play a central role in achieving the mission objectives means that meeting the magnetic cleanliness design goal of the spacecrafts is an important requirement for the success of the mission.

13.1. INTRODUCTION

It is the role of the EMC Review Board to coordinate all relevant activities to ensure that the magnetic cleanliness design goal is met or bettered.

The objective of this guide is to provide indications on how to achieve acceptable magnetic cleanliness levels. It considers the effects of hard and soft magnetic parts and materials, and those of stray fields, resulting from currents in the spacecraft and payload electrical systems. Permanent fields can be controlled to some extent by careful positioning of magnetic components (Relays, for example can be positioned in self-compensating way), however it is not always possible because of mechanical restrictions to position optimally some of the major contributing sources which have known (thrusters, valves , Reaction Wheels for example)or predictable magnetic characteristics.

In addition , magnetized soft magnetic material in the vicinity of a permanent magnet may produce much less stable contributions than the magnet itself. Therefore parts and materials to be used in the spacecraft must be selected so as to minimize their contribution to the background magnetic field.

In considering fields from magnetized parts it is important to recognize the difference between hard and soft magnetic materials. Whereas the contribution of hard magnetic materials can be expected to be quite stable, magnetic field of soft material may vary appreciably, particularly during launch and in response to varying electrical activity in the spacecraft, i.e. in response to stray fields from currents. The hard field background due to permanent magnets, for instance , can be determined by a combination of ground testing, modeling and in-flight calibration and is, if not large, therefore often of lesser concern . The strongly varying contribution of soft materials, on the other hand, must be kept low by strictly limiting the use of such materials on the spacecraft.

Control of stray fields from currents (including solar panels) must also be considered during design phase to minimize their effects on the magnetometers. As with permanent fields, positioning is important and must be considered, along with harness layout, shielding, etc. In addition, full consideration must be given to acceptable EMC/EMI design practices which will satisfy the spacecraft EMC specifications. This document deals with DC and very low frequency disturbances only, for AC interferences the demands are in general much higher.

Previous spacecraft projects like HELIOS A,B, GEOS, ISEEE, GIOTTO, TETHER, ULYSSES,Mars-96, CLUSTER, and ROSETTA have conducted extensive and successful magnetic cleanliness programmes to ensure the success of the mission. The material presented here is based on this previous experience. This document is intended as a guide for magnetic cleanliness design of spacecrafts, and presents information on various aspects of the parts, materials and equipment to be used. One of its objectives, in particular, is to assist experimenters with design of magnetically cleaner experiments to minimize residual fields and thus to help achieve the mission aims of the spacecraft under discussion.

13.2 Design Approach

13.2.1 General

To meet the mission requirements, considerations of magnetic cleanliness must have a high priority during the design phase, selection and positioning of flight components and hardware need to be carefully evaluated for possible magnetic contamination. The selected design of hardware in the integrated spacecraft shall of course not conflict with other performance criteria defined in the applicable specifications. This may relate to allocated weight,power,other EMC design considerations and environmental considerations (radiation, thermal, vibration and shock). Also, reliability, feasibility,available experience, available materials, as well as cost and schedules remain important considerations.

13.2.2 Approach to Minimizing Fields from Magnetic Materials

For the reduction and control of magnetized material field contamination levels, the major areas of concern are as follows:

1. component and piece –part evaluation; selection of optimum items.

2. Magnetic test of all electronic parts and mechanical components.

3. Optimum parts layout for circuit boards

4. Optimum component or assembly orientation for mutual compensation.

5. Provisions for the safe performance of components during deperming (degaussing) operations.

6. DC field compensation.

13.2. DESIGN APPROACH

13.2.3 Approach to Minimizing Stray Fields from Electrical Currents

To comply with the requirements of the experimenters for minimizing stray fields from currents, the following areas shall be considered (noting that all stray fields above 10Hz are considered in the realm of EMI/EMC):

1. Design of spacecraft wiring and power distribution system, with special attention to

2. Bonding, shielding and grounding.

3. The use of paired, twisted leads wherever possible. No supply current shall flow into the spacecraft structure.

4. Equipment location within the spacecraft

5. Elimination of current loops throughout the spacecraft

13.2.4 Approach to Minimizing Fields from Eddy Currents

1. Skin, electrostatic shields, thermal shields, etc. should be made as thin as possible and as high resistive as possible

2. All conductive structural loops should be avoided

13.2.5 Magnetic Moment Management , Approach to Hardware Analysis

For a number of important magnetic material field sources, an approach to achieving good magnetic cleanliness levels may be the following: the magnetometer outboard sensor (as used on many spacecrafts)is extended at a distance of R = 2,75 m (on a radial 2m boom) from the center of gravity of the spacecraft having a mass M = 110kg.The total tolerable DC field perturbation for example is $\Delta B = 10nT$ at the outboard sensor which allows for a total magnetic moment referred to the spacecraft center of
m_m =1040mAm^2*

(for definitions and units, see Appendix A)
*(This is defined as the magnetic moment of the spacecraft in the DE-PERM Status, typically for PERM Status about 50% and for Stray Field

Status about 30% increment are allocated.) Distributing this magnetic moment evenly among the experiments and the S/C subsystems according to mass then allows

$1040/110 =$ **9,4 mAm² /kg**

for all systems and components (special care must be taken for small units having a mass below 1kg and for the solar panel).

A more sophisticated calculation and distribution taking into account the built in positions of all units (as far as known) and its different distances wrt. the outboard magnetometer is based on a special S/W called ALLIST as part of the MAGNET dipole modeling software. This should be applied in a later project phase (ask the author,K.Mehlem or A.Stadelmann).

It is recommended that the EMC board will approve a list of acceptable magnetic moments allocated to the individual experiments and subsystems.

(A list of more than 300 units measured magnetic moments exists from previous spacecrafts for comparison)

This process will in fact aim at a dialogue between the EMC board and the hardware suppliers, where the suppliers are asked to report their expected magnetic moments (magnetic mapping in one of the small MCFs , Mobile Coil Facilities, is recommended) as early as possible to the EMC board. This also preassumes that all persons involved in flight hardware keep the amount of magnetic moment in their units at or below the allocated amount.

Any unused magnetic moment will be used to readjust the allocations as well as any too large magnetic moment will start a collaborative effort with magnetometer team, EMC board and the unit supplier to propose a solution. It is very important that all subsystems and units be surveyed for magnetic field sources as early as possible to allow for flexibility in available solutions.

13.2.6 Shielding Problems

It is not desirable to intentionally add magnetic material to the spacecraft. However, in some cases it is unavoidable. Magnetic shielding material should be used carefully in minimum amounts. It should only be used to prevent from varying magnetic fields from sources such as stepper motor and stepping electromagnets that are sometimes used in mass spectrometers. The shielding material should be used close to the source of varying magnetic field and integrated into the structure of the device. For example, it requires far less material to directly shield a motor than it would be to shield the entire experiment box that contains the motor.

Many energetic particle experiments use magnets to focus or to sweep out undesired particles. These should not be shielded. The design of the magnets should be such that there is a minimum of fringing field. This can be accomplished by careful magnet matching and using a properly designed yoke to close the magnetic path of the magnets. If there is a net residual field from such a magnet, the Review Board may request that compensation magnets be attached to the experiment to null this field. Shielding and compensation should be added only at the direction of the EMC Review Board.

13.2.7 The Approval Process

1. If you suspect that a problem may exist first test the parts. In case of varying fields a facility with a fluxgate magnetometer may be required.

2. Contact the EMC Review Board and report the results of your test. The EMC Review Board will assist you in testing and finding solutions to the problem.

3. If shielding or compensation is required, file a formal request for deviation of the specification.

4. The EMC Review Board will make the appropriate disposition of the request for deviation.

13.3 Achievement of Magnetic Cleanliness

As stated earlier, there are four sources of magnetic fields; materials which can be or have been, permanently magnetized, soft magnetic materials, currents in the spacecraft electrical subsystems which produce magnetic stray fields and eddy currents induced by electrical fields in conducting structural loops or varying fields inducing currents in conductive structural loops.

Materials are defined as all structural, mechanical and electrical material and hardware. Manufacturers of components and those handling and mounting parts should be aware that otherwise non-magnetic items can be contaminated by magnetized tools or transferred by non magnetic tool previously used on a magnetic material. The importance of using non-magnetic tools (or at least degaussed tools) increases the closer to the magnetometer sensor the tools are used, but good working practice includes the use of non-magnetic (or degaussed) tools everywhere. The danger of transferring magnetized particles to non-magnetic materials again stresses the importance of testing all parts and tools before use.

13.3.1 Permanent Fields and Fields from Soft Magnetic Materials

All materials considered for use in the spacecrafts and during AIV activities on the spacecraft, even those thought to be non-magnetic, must be carefully tested prior to their use.

Materials

General

Aluminium, fiberglass, carbon fiber, magnesium and titanium are all non-magnetic.

These are among the most desirable materials for use in structures. Steel or other magnetic materials should not be used in the structure or mechanical hardware.

All materials considered for use in the spacecraft, if not known to be satisfactory, must be carefully tested prior to their use.

The documents referred to give sample lists of normally non-magnetic materials. Being on these lists is, however, no guarantee that samples bought from a supplier will be non-magnetic(for example stainless steel tanks), impurities and changes in the manufacturing process*, etc. may result in the material being magnetic. All samples, etc. and items must be tested. As an example I refer to the delicate goldcoating of an Aluminium box (GIOTTO Transponder), this always first has to be coated by Nickel, this however is magnetic if the layer is too thick(several μ only) , therefore read handling manual carefully for this.

Another example is the magnetized INCONEL for propellant tanks, INCONEL is non-magnetic however the machining sometimes causes terrible magnetization .

1. Welding Wire

    Ordinary nickel welding wire used for inter-connections between components in welded modules is highly magnetic. As such, it should not be used on the TBD project. A Nickel-Copper alloy should be considered for welding. This alloy is composed of 78% Copper and 22% Nickel. Although a significant fraction of alloy is Nickel, the alloy remains non-magnetic through welding, heat treatment, vibration and all environmental testing. This alloy is easily welded and considered as reliable as Nickel.

2. Plastics and Epoxies

   Plastics and epoxies are not magnetic in themselves but some fillers used are magnetic. Care should be exercised in the use of red and black fillers especially since these may contain iron in various oxide or metallic forms as coloring agents. Problems are not expected when other fillers are used, such as white or green. In all cases, samples of epoxy or plastic used should be tested (sniffed) to verify its magnetic acceptability. (high temperature glue or epoxies often use magnetic fillers!)

3. **Electronic Parts**

4. Resistors

   Non-magnetic metal film and carbon composition resistors can easily be obtained if non-magnetic lead materials are specified. Care should be taken in the choice of metal film resistors to avoid the use of those having a spiral or helical pattern if passing currents will produce large stray magnetic fields.

5. Capacitors

   Tantalum Capacitors:

   Non-magnetic tantalum capacitors are difficult to find (the use of Mini-tan caps is recommended). The source of magnetic field is in the glass to metal seal and the magnetic lead material commonly employed. However, non-magnetic sintered tantalum, electrolytic slug capacitors and non-magnetic electrolytic tantalum foil capacitors are available with non-magnetic lead material.

   The magnetic fields of other types of tantalum capacitors can be reduced somewhat by using the non-magnetic Nickel-Copper alloy leads in place of the highly magnetic Kovar leads with glass- to- metal seal.

   Other Capacitors:

   The following types of capacitors are available in non-magnetic forms:

   - Fixed Silver MICA Dielectric Dipped Coating
   - Fixed Ceramic, Dielectric, Filter, Feedthrough
   - Fixed Glass Dielectric

- Fixed Ceramic, Dielectric (general purpose)
- Fixed, MICA Dielectric
- Variable, Piston
-

(a) Crystals

Packing of crystals in titanium or plated brass is desirable since this type of packing is non-magnetic.

(b) Relays

Since more reliable mechanisms depend on magnetic actuation for switching and permanent magnets for latching, the use of relays should be limited to only the most critical functions which cannot be handled by solid state switching.

The magnetic field of the permanent magnet in a latching relay can be minimized through the choice of the smallest, least magnetic relay adequate for the task. Further reduction can be accomplished by the addition of a small permanent compensating magnet, sized and positioned on the relay case so that its magnetic field partially cancels the magnetic field of the magnet internal to the relay.*

In case of banks of (at least two) relays, these should be arranged in back-to-back configuration to achieve maximum mutual compensation.

* The compensation of the relay internal magnet by an external magnet on the relay case must be tested carefully because of dangerous oscillations in case of too close fixation of the external magnet.(see ROSETTA, CLUSTER Thrusters and Valves compensation).

(c) Wire and Cables

The following precautions are necessary to ensure that non-magnetic wire and coaxial cable are used

(d) No wire with plated steel conductors should be used.

(e) No shielded wire with braided steel mesh shielding should be used.*

* R/F cables often use strong magnetic springs for fixation (see GIOTTO)

(f) Connectors

It is necessary to specify non-magnetic connectors like Cannon-NMB and AMP Series 109

(watch long lead time). Care must be taken when ordering connectors containing springs and bayonet type locking mechanisms. A number of R/F connectors contain Kovar glass-to metal seals. Test (sniff) should always be made on samples of any connector before placing the final order.

(g) Transistors

When using single transistors, non-magnetic packages should be given preference. Samples should be tested before ordering.

(h) Diodes

Although most diodes are commonly available in glass packaging employ large , highly magnetic glass-to-metal seals and lead materials, several types of more acceptable packages are also available. Among them are the " DO-7" package and the "Adam"package. At times it may be necessary to use diodes which are available only with the highly magnetic lead materials, but then the leads should be cut as short as possible.

(i) Integrated Circuits

By minimizing the lead length to the smallest practical value, the magnetic field of integrated circuits can be reduced. Careful layout, avoiding the addition of small magnetic contributions, should be used. The resultant field of assemblies can also be minimized by back-to-back mounting of electronic boards.

(j) Ferrite Cores and Pulse Transformers

General , ferrite cores exhibit less magnetic field after magnetizing than powered iron cores of similar size. The size appears to be the only factor in determining the magnetic moment retained by the core. Since the powered iron cores retain higher magnetic moments after exposure to a magnetic field, the use of powered iron cores should be avoided wherever possible.

Many small pulse transformers are acceptable for use. The size of the core or bobbin determines the magnetic field retained after exposure to a magnetizing field.

Small cores or bobbins wound with non-magnetic copper wire are preferred.

(k) Power Transformers, Chokes and Inductors

All magnetic components must be carefully wound on toroidal cores with non-magnetic copper wire (using bifilar winding techniques when possible and appropriate). In this manner, the permanent and variable stray fields of the magnetic components are minimized.

13.3.2 Stray Fields Produced by Electric Currents

Guidelines

Many techniques have been identified for minimizing the stray fields produced by currents within the spacecraft, among which the following are of particular importance:

(a) Leads carrying appreciable current (greater than 1mA) must be twisted with the return lead, such that the net current in the twisted wires is as near zero as possible. Even if it is not possible to achieve a null net current, parallel cancelation is still desirable.

(b) In wiring through connectors all leads should be kept as close as possible to their return to obtain the best possible self-cancelation through the connector.

(c) All connectors should be placed in one particular area of the assembly, this area being as small as possible.

(d) All power wiring throughout the spacecraft requires twisted cabling

(e) Extreme caution must be exercised to avoid circulating ground loops through the structure

13.3. ACHIEVEMENT OF MAGNETIC CLEANLINESS

(f) Heater foils and elements must be of non-inductive design. Precautions 1, 2 and 3 aim at reducing the possible area of the current loops in the assemblies and spacecraft harness. A large moment can be built up by summing many relatively small moments. Thus, the basic approach is to minimize even the smallest loops within the assemblies and harness.

Precautions 1 and 2 are especially necessary when dealing with assemblies in Power Supplies. The current carrying leads internal to the converter assemblies must be carefully routed and twisted wherever possible to provide cancelation of stray fields.

All wiring in the spacecraft harness carrying more than 1mA should be twisted with the return, with the number of twists per unit length determined by the gauge of the wire used. The twisting must be tight enough to prevent "birdcaging" of the wires but not so tight that the wires will be twisted into solenoids. Typically, the number of **twists ranges from 1 to 0.3 turns per cm** Precautions 4 is a restatement of a, specifically dealing with the power distribution leads. Precaution e suggests a single – point power grounding system to avoid uncontrolled, circulating ground loops in the equipment platform. The prime danger is that the ground loops cannot be accurately calculated in that they may not be completely controlled. In addition, they are very hard to determine experimentally prior to the magnetic testing of the completed, integrated spacecraft.**

** MARS Observer that got lost after one year flight shortly before reaching MARS used two separated ground bars producing several 10nT noise at the 6m boom tip with the magnetometers.

Stray Fields of Parts with Magnetic Cores

In general, the construction of magnetically acceptable power transformers, chokes and inductors is a problem. As previously discussed, these parts can be constructed with very low permanent core fields. However the variable stray fields from currents associated with them can be quite large and care must be taken to reduce them to an acceptable level. Although most currents in the transformers and inductors are high frequency, DC offsets are usually present in at least one set of windings and large DC currents are always present in the chokes. Studies have been made to determine the factors which significantly affect the stray magnetic fields of the parts when direct current is applied. The results of these studies may be summarized as follows:

(a) The windings, whether they be few or many, must be uniformly spaced around the toroid.

(b) The windings must be wound very tightly to the core to reduce air gaps. The use of multifilar magnet wire is recommended. Also, it was found experimentally that the stray field is reduced if the core is rolled on a hard surface after winding, thus improving the fit of the windings to the core.

As an added precaution, even after taking these precautions, extra parts can be ordered and only the best ones (after tests) should be selected for use.

Two sets of magnetic screenings can be made to find the best parts: One after winding but before potting (and the subsequent reliability testing) , one at incoming inspection of the finished part. A pre-potting test will predict the rejection of those obviously unacceptable, and will determine which are the best of the lot and which are marginally acceptable.

Solar Array

The stray fields of solar arrays can be minimized by the technique of **"backwiring"**. In backwiring, the return wire from each string of solar cell modules is returned behind the modules of that particular string and carefully routed along a line just behind the centreline of the modules.

The advantages of backwiring are obvious. Each string and module of the a string is self-canceling and does not depend on the magnetic field of adjacent module or string for cancelation. Thus, if a module fails during flight, the current in both the string and return drops to zero simultaneously, leaving no uncompensated currents in the array. Similarly , if the current level is reduced in a string through the loss of cells, the current in return wire (which is the same current) is reduced so that there is no imbalance.***

*** There was a very strong magnetic field perturbation observed on DSP 1 (Chinese Double Star) because of no backwiring at all and second

13.3. ACHIEVEMENT OF MAGNETIC CLEANLINESS

a very strong spin modulation because of separating power line and return wire thus having two big circular loops (magnetic moments).

On DSP 2 this was reduced using backwiring.

There are special 3-D graphical programs developed for calculation of the DSP solar panel fields at the magnetometer position also during spin.

Another 3-D magnetic field program for the large ROSETTA panels has been developed showing the magnetic fields at specified points (magnetometer) also during spin and also introducing loss of solar cells or strings.

A third 3-D magnetic field program has been developed for MARS Netlander circular solar petals. This program can optimize the cell distribution and string wiring on a petal.

These software packages have been developed by
H.P.Brunke,GEONUMERIX,Braunschweig

On Helios A a short between Solar panel return line and structure was observed during system test producing several 10 nT magnetic field at the magnetometer .However this only could be detected because of the possibility to power the solar panel artificially via the backwiring.

Special Problem Assemblies

Several assemblies on TBD spacecraft can be classified as "special problem areas". A few notes are already included here for guidance.

- moving mechanical assemblies, stepper motors, etc can cause special problems, and therefore require special attention for a magnetically clean design. In particular, any use of motors should be cleared with the EMC Board

- Batteries can be a major source of background field (charged and uncharged). However, careful design and compensation techniques have been developed to cancel the external field, for instance on ULYSSES and GIOTTO missions.

- Antenna RF components use magnetically hard material (springs, magnets, gyrators, circulators). These have to be inspected and sniffed very carefully, and degaussed piece by piece.

- Travelling Wave Tubes are a known source of background field. Their contribution must be minimized by positioning for self-compensation, and or by using compensating magnets.

- Reaction wheels are also a known source (DC/AC) of magnetic fields, their positioning wrt. the magnetometer should be early discussed.

- Thrusters, Valves caused big magnetic problems on CLUSTER, their compensation with more than one magnet per unit caused hard work during testing and gluing the right magnet in the right direction on given small areas on the unit.

If possible the selection and testing of Thrusters and Valves having small internal magnets only should be made as early as possible.

- The decision on common power converter frequency should be discussed very early.

- The solar panel design including backwiring must be tested during design phase and a possibility to test the solar panels within the system test must be planned for.

- All experiments are generally sources of magnetic fields (including focussing magnets), their design and early testing must be planned.

- Propellant tanks must be sniffed at the developing industry for non-magnetic material used and possible magnetization during machining

Fields form Eddy Currents in Structural Components

The magnetic disturbance from eddy currents induced by the ambient strongly varying Earth Field in a rotating conducting body may be considerable, particularly in other high planetary fields' environment and where highly conductive closed structural loops are involved.

A solid Copper ring 2m in diameter and $2cm^2$ cross sectional area, spinning at a rate of $3rad/s$ in a $1000nT$ field will have a magnetic moment equal to 27.3 mAm^2 from the induced eddy current. This is about the typical magnetic moment allocated to a 3kg experiment.

Conducting closed loops or closed shells in the spacecraft may thus present a magnetic cleanliness problem to be taken seriously. The guidelines to minimize this include the following:

(a) Where not prohibited by mechanical restraints, non-conducting or high resistive materials should be chosen for structural parts.

(b) Conductive structural parts should be designed in a way to avoid closed conducting loops.

(c) Any conducting loop should be disrupted electrically, e.g. , by suitable insertion of a low-or-non-conductive section.

13.4 Magnetic testing of Spacecraft: Flight Hardware

It is recommended and stated in the Guidelines that all materials used in the spacecraft hardware as well as the tools used close to the spacecraft during AIV activities should be magnetically sniffed.

Complete assemblies and units for which a magnetic specification and allocated magnetic moment are issued will be tested as units to demonstrate that the specification and allocation are met. Furthermore the units will also be magnetically mapped for determining the input to the spacecraft magnetic model. It is recommended that already the EM units should be mapped for possible changes to be made for the Flight Units.

13.4.1 Testing at Assembly and Unit Level

In order to verify that the requirements are met , the magnetic signature of the experiment and other S/C units has to be determined according to a standard procedure. This procedure shall permit control of the unit levels of fields such as:

(a) to demonstrate that each experiment complies with the allocation of magnetic field emissions given in the EID for experiments and in the EMC Compatibility Requirements Specification given for the spacecraft subsystems.

(b) To provide data on the magnetic properties of each unit in order to be used in the magnetic analysis of the complete spacecraft using the ESA Magnetic Analysis Programme (MAGNET)

The magnetic field tests shall be carried out on all EM experiment units, using an appropriate facility. 5 of those small so called Mobile Coil Facilities MCF have been developed by ESA/ESTEC (built by Pfeil), they have been used for more than 25 years for GIOTTO, ULYSSES, TETHER, CLUSTER, CASSINI, and ROSETTA, two MCF's exist at ESTEC (currently used for LISA at ESTEC respectively at EADS Stevenage), another MCF facility exists at IC-London, currently (2009-2010)used for SWARM at EADS Friedrichshafen, another MCF exists at TU-BS and one at IWF-Graz. The facilities have been mostly upgraded with new Hardware and Software (MAGNET) by PFEIL-TRAWID company (for questions regarding the hardware or software ask:
G. Musmann , e-mail: guenter.musmann@freenet.de
or D. Pfeil ,e-mail: Pfeil-Trawid-0001@t-online.de,
or K.Mehlem,e-mail:Klaus.Mehlem@web.de.)
H.P.Brunke,GEONUMERIX,Braunschweig,e-mail: H.Brunke@geonumerix.de

The standard test procedure developed includes a test sequence, each step is followed by a single plane $360°rotational\ mapping\ and\ analysis$:

1.Incoming;

2. Deperm with 40-50Gauss, 3Hz ;

3.Stray field test (means power on status);

4.Perm Field susceptibility (1,5-3 Gauss Perm);

5.Final Deperm;

6.optionally Induced test

in case of soft magnetic material in the unit.

The results are on-line printed as standard Test Report document.

(Typically such unit test takes 6-8 hours test including test report.)

At the end of this magnetic test a decision has to be made and reported whether the unit can be integrated as it is or whether a unit compensation has to be made or whether the unit has to be refurbished.

13.4. MAGNETIC TESTING OF SPACECRAFT: FLIGHT HARDWARE

During integration activities of the Flight spacecrafts one of the MCF's should be installed close to the spacecrafts so that all units could be tested according to the above procedure prior to integration.

Finally after integration of all units a System Magnetic Test has to be planned for, including a sniff test along the magnetometer boom to look for hot spots. (Cluster Super Insulation Foil staples were strongly magnetic close to magnetometer Sensor.)

In many cases a magnetic cleanliness requirement has to be fulfilled. In the case of interplanetary missions the interplanetary field encountered is typically

$$0.1 - 20nT$$

between Pluto and Mercury. Typical DC-magnetic cleanliness specifications are in the order of

$$0.1 < b_{Spec} < 1.0nT$$

($Ulysses : 0.1nT, Cluster : 0.25nT$) at the magnetometer location (Specification Point). The verification of the cleanliness specification has to done on the ground in the Earth field

$$(50000nT)$$

in coil systems. Large facilities achieve a zero drift stability in the order of

$$0.1 - 0.3nT$$

and in a volume of homogeneity with a radius of at most $\Phi= \tilde{4}$m. S/C with deployed magnetometer booms, for instance 6.45m (Ulysses), exceed this volume. Even if the spacecraft fits the volume, due to the precision of the coil facility, the verification of the magnetic cleanliness specification by direct measurements is not possible. Simple extrapolations of near-field measurements with suitable signal-to-noise ratio to far-field distances are very problematic due the higher orders appearing in the near-field measurements. An additional step is therefore required in order to generate precise estimates of the non-measurable far-field, in particular at the Specification Point. This becomes possible by deriving a numerical model of the magnetic potential on the basis of near-field measurements and by simply calculating the far-field.

The modeling method is based on the postulate that any magnetic object can be represented by a finite set of dipoles. The so-called **Multiple Dipole Model (MDM), developed by K.Mehlem of ESA** (GANEW, MDM and MAGNET software), is obtained by optimizing the individual dipole positions and moments in the sense of a least-square fit of the calculated near-field to the measured one. The measurements are obtained in the classical rotational mode. As a result of this system test it has to be decided whether a compensation using magnets is recommended or not. This can be done as so called Point compensation, means to calculate and select a magnet and its location, direction and rotation angle on the spacecraft such that it compensates the residual S/C field at the magnetometer. (Unit compensation is the other option; means each unit with a magnetic moment far above the allocation has to be compensated, but this has to be discussed individually) It is recommended that for the compensation purpose the Magnetostatic Evaluation Tool from MAGNET II software , so called Compensation software is used. This software selects from a series of existing magnets the appropriate best choice, calculates it location on a given plane in the spacecraft and its directional angles.

13.5 Appendix A

13.5.1 Relations to other units :

$1 nT = 1\ \gamma = 10^{-5}$ Gauss (G)

$1 mAm^2 = 1\ G.cm^3 = 1$ " c.g.s" $= 1$ pole-cm $= 1$ "emu"
$= 0,1 nT\ m^3$

The Earth has a magnetic dipole moment equal to :

$m_E = 7,87 \times 10^{22}\ Am^2$

(International Geomagnetic Reference Field 1985, the dipolar field part is reduced by about 6% within the last 100 years !)

A small pocket magnet may have a magnetic moment in the range :

$m_m = 0,2 - 2\ Am^2$

A small coil of 100 turns, with a winding area equal to 2 cm^2 and a current of 100mA has a magnetic moment equal to:

$m_m = 100 \times 100\ mA \times 2\ cm^2 = 2 \times 10^{-3}\ Am^2$

At R = 1m distance along the coil axis the magnetic field is: (1. Gauss'sche Position)

$B = 2\ m_m/R^3$

$B = 2/\ 10^7 \times (\ 2 \times 10^{-3})/1^3 = 0,4 nT$

At the equator

this field is only half the polar field: (2. Gauss'sche Position)

$B = m_m\ /R^3$

13.6 Appendix B

13.6.1 GENERAL GUIDELINES FOR THE DESIGN OF MAGNETIC CIRCUIT FOR MAGNETICALLY CLEAN SPACECRAFT

General Approach: the basic objectives of the design of magnetic circuits for spacecrafts instruments are generally the following:

(a) Obtain the desired flux density at the gap (a)

(b) Minimize weight and volume(b)

(c) Minimize stray and leakage flux(c)

(d) Maximize stability (temperature, time, B)(d)

In the following discussion we will emphasize mainly objective (c)

Obviously, objectives (a) and (b) imply by definition objective (c). The minimization of stray and leakage fluxes is generally accomplished by reducing the reluctance of the magnetic circuit to a minimum by the use of pole pieces and yokes attached to the permanent magnets. This technique leads directly to the accomplishment of (a) and (b) above. When dealing with magnetically clean spacecraft objective (c) needs to be examined in more detail since very small leakage fluxes which have practically no effect on the gap flux density can have very large effects on the far field generated by the magnet- yoke-pole pieces assembly, particularly if the maximum allowable field is of the order of a few nT at, let's say, 1 meter distance . In general, the far field will be dipolar in nature for distances greater than 5-10 L, where L is the characteristic dimension of the magnet assembly.

Other factors that must be considered are grouped in objective (d).That is how stable is the leakage flux to changes in environmental conditions, including external magnetic fields.

Discussion: To minimize weight and volume a high saturation flux density material of high permeability is chosen for the construction of the yoke and pole pieces. In general, these materials lose their magnetic properties when stressed during machining or other forming operations and need to be annealed in a dry hydrogen atmosphere. The high permeability of the material presents a problem from the point of view of susceptibility of the stray flux

13.6. APPENDIX B

to external magnetic fields. Very small changes in external field lead to large changes in stray fields which is highly undesirable.

Fortunately, if we dimension our yoke and pole pieces such that they are working near saturation flux density levels (the knee of the B-H curve) not only do we reduce susceptibility to external fields (low incremental permeability but also we minimize the area of material required to carry the desired flux and thus the mass of the assembly (assuming constant length).

A material that has been used with success in the machining of yokes and pole pieces is "CARPENTER 49-FM " or Mu-metal designed for improved machinability.. The saturation flux density for 49-FM is approximately 15,000 Gauss while the "knee" of the B-H curve occurs approximately at 8000 Gauss. This is a good starting point for design purposes. Although not absolutely necessary if machined with care, it is recommended that the machined parts be annealed as recommended in the literature.

In general, a yoke-pole –piece assembly will be composed of a number of parts which must fit together. Wherever a joint occurs the surfaces should maintain contact over the entire available area to minimize fringing and stray fields. This is generally accomplished by polishing and lapping of the surfaces to mirror finishes and the use of a large number of attachment devices (screws) along the joint to distribute uniformly the assembly stresses. The screws should be made of non-magnetic materials (Titanium, brass, etc) to avoid creating local field concentrations and resulting large stray fields. As a general rule, the magnetic assembly should be " smooth", i.e. without protrusions, indentations, feet, etc. which lead to highly localized field gradients and resulting stray fields.

Only the minimum number of openings should be provided for access to the high field gap region. This implies that the design in general will be a box with slotted apertures for entrance and exit of charged particles. These apertures should be as small as possible and located in regions of low field gradients. Since each aperture is a source of leakage it is possible to minimize the stray field by shaping the aperture surface such as to concentrate the leakage flux inside the magnetic assembly rather than the outside.

The usual development cycle for a magnetic assembly with low stray field calls for the construction of a prototype which can be measured locally with a HALL-probe magnetometer to determine the size and location of leakage flux sources. These are then corrected by techniques described here. Further tests can be performed in a magnetic test facility to determine the far field characteristics. An analytic model is useful in the early stages of the design process but since the leakage flux is always associated with small variations in structural properties a bench test is always required to certify the design and to develop the proper assembly techniques.

References chapter 13

(a) Lundsten, R. H., and D.I. Gordon,: Metallic Materials for Nonmagnetic Spacecraft, Properties and Measurement Methods. Proceeding of the Magnetics Workshop March 30-April 1,1965,edited by J.G.Bastow, NASA JPL Techn.Memorandum No.33-216,pp.317-329,1965

(b) Musmann, G.,: Design Guide for Magnetic Cleanliness Control, internal paper GIOTTO, 1982

(c) GIOTTO: Parts and materials List for magnetically Cleaner Spacecraft, ESA

(d) Musmann,G. Balogh,A. Schwingenschuh,K., CLUSTER: Guidelines for Magnetic Cleanliness on the CLUSTER Spacecraft, prepared by Cluster Magnetometer Team;1986

(e) Musmann, G.,:Problems with Magnetic Field Measurements on Spacecrafts, Dt.hydrogr.Z.41, pp 265-276,1989

(f) Primdahl,F.,A Pedestrians Approach to Magnetic Cleanliness, DRI 2-90, Danish Space Research Institute, Internal Report, 1990

(g) Christy, J.R.,: Magnetic Classification of Metallic Materials, ibid., pp 345-357

(h) Mehlem, K., Musmann, G.; ESA-ESTEC ,2002 DSP DEEP SPACE ONE : Magnetic Cleanliness.

13.6. APPENDIX B

(i) Mehlem, K: Multiple magnetic dipole modeling and field prediction of satellites, IEEE Trans on Magnetics, September 1978.

(j) Mehlem, K: Ulysses RTG F3 magnetic compensation, ESA/ESTEC Working Paper 1537, March 1989.

(k) Mehlem, K: Cassini RTG F5, F2, F7, F6 magnetic analysis report, ESA/ESTEC-JPL internal working paper, 1996/97.

(l) Mehlem, K:Multiple magnetic dipole modeling and field prediction of satellites ESA/ESTEC/TMM, Noordwijk, Holland; IEEE Transactions on Magnetics,Publication Date: Sep 1978 Volume:14,Issue:5

(m) Kuhnke,F.,M.Menvielle,G.Musmann,J.F.Karczewski,
H.Kügler,C.Cavoit and P.Schibler
The Optimism/Mag Mars-96 experiment:Magnetic measurements onboard landers and related magnetic cleanliness program.Planet.Space Sci.,Vol.46,No.6/7,1998

(n) Mehlem,K.,Narvaez,P; ESA-ESTEC,Noordwijk; JPL,Pasadena; Magnetostatic Cleanliness of the Radioisotope Thermoelectric Generators (RTGs) of Cassini, IEEE International Symposium on Electromagnetic Compatibility,1999,Seattle.

www.ingramcontent.com/pod-product-compliance
Lightning Source LLC
Chambersburg PA
CBHW082322220526
45470CB00008B/2378